DREAD THE FRED

JOYCE C. RAGLAND

Paperback-Press
an imprint of A & S Publishing
A & S Holmes, Inc.

Cover Design by Phillip Foust
Cover Art by Andi Osiek
Author photo by Nathan Howard

ISBN: 0-9911805-0X
ISBN-13: 978-0-9911805-0-9

DISCLAIMER

Before writing this book, the author interviewed as many people involved with the Robotics Club, and the R-I district, as would sit down and talk with her. Any errors in fact are entirely the author's responsibility, and never meant to be detrimental to any person or any locale. Conway High School is the author's alma mater, and she is proud to have that history.

CONTENTS

DEDICATION

This book is dedicated to all those who supported the Conway
Robotics Club, especially during the FRED year

ROBOTICS CLUB yearbook pose 2009-2010

ACKNOWLEDGEMENTS

I thank everyone who took the time to talk with me about the FRED year, especially Paul Coryell, Phillip Foust, Ceira (Gisselbeck) Fields, Shyanne Witt, Grant Rumfelt, Lloyd Oberbeck, Jr., Shane Sell, Broghan Fields, Bob Gibson, Chris Berger, Billy Coyle, Robert Coryell, Pat Cunningham, Teresa Rumfelt, Ruby and Lloyd Oberbeck, Sr.

Big thanks to Kathy Garnsey for editing the final draft.

A super-sized thank you to Sharon Kizziah Holmes for believing in this book and bringing it to publication.

Joyce Ragland

ONE

PAUL

We didn't mean to almost kill someone during after school robotics work, but Gibby sometimes thought we did. We knew to hide behind the big sander in the shop the time we really ratcheted up the air compressor and ran a little experiment and besides, no one else was in the shop that evening. Those occasional experiments happened only after we'd been working on the robot for several hours and became a little slap happy. Phillip would say, "I wonder what would happen if…"

I'm Paul Coryell, the chief engineer for the robot. My best friend since kindergarten, Phillip Foust, assisted my creative endeavors. Most of the time we were serious but when we got crazy tired, things could get interesting. We were high school juniors and, well, creative.

Mr. Gibson—Gibby, our fav math teacher and the Robotics Club sponsor, was grading math students' papers in the next room so he was close, but not actually in the shop and the doors were wide open. He was there after school because students couldn't be using shop equipment without a teacher close by.

That day we discovered some of the shop kids—slackers who took shop classes for a lark—had broken the nozzle end of the air compressor hose. It had the trigger, but the air was at twice the normal volume and pressure. We wondered what that much air pressure might do. Phillip grabbed a piece of PVC pipe from our parts kit and we duct taped it to the end of the compressor hose. I got a steel rod from the kit and stuck it in the pipe and we let the tank's air pressure go up a bit then let go of the trigger. The pipe just sort of flopped out with no velocity, no force.

"Well that was not impressive." Phillip said.

I said, "What if we had something to build up the pressure..." I looked around and saw the towel dispenser. "Let's try wads of paper towels."

"Do it," Phillip said. "Aim toward that hunk of board on the back wall. That's what, twenty feet away?"

So we stuck the paper towel wadding into the PVC tube still taped to the hose, and then we put in the steel rod. We had at least 150 psi pressure built up, way more than normal. We hid behind the big sander in case of a ricochet. We didn't really expect one but still, you never knew. I held the air tube and Phillip turned on the air compressor at the wall then crouched down as I shot the trigger.

BAM!!

Mr. Gibson came running in and yelled, "What the hell?"

Phillip and I were laughing so hard our sides almost split but we tried to calm down.

We went over and looked at the big thick board. It had a huge gouge in it.

I said, "It worked!"

"We had our safety glasses on," Phillip said with a totally innocent face.

"You idiots could get killed one of these days!" Gibby said, then laughed.

Whew!

We worked really hard after that for the rest of that day. Comic relief works.

Mr. Gibson was a great teacher–the best mathematics teacher I'd had, and we got along on a special level. He'd tease Phil and me and we'd tease back, but only after school, not so much in class. I knew my boundaries.

The air compressor situation happened the year before, but Phillip and I talked about it as we started the new school year. You never knew what equipment the slackers would've broken and for building the robot after school and on weekends, we needed all the equipment to work.

Paul & Phillip

Gibby - Grrrr

TWO

+

CEIRA

"Oh crap!" burst out of my mouth when Mr. Gibson told me that Mrs. Day wouldn't be involved with Robotics Club that year. I didn't yell or stomp my feet or anything like that, but it shocked me. Big time. We had to have the mock corporation or the robot couldn't enter the BEST robotics competition so without a teacher's help, I'd be all on my own.

"What's going on?" was probably the second thing I said. I'd gone into a zone of disbelief. Mr. Gibson said something about Ms. Day having some health problems.

"Can you do it without her?" He jarred me into consciousness. "Head up the table display and the faux corporation components? I won't have much time to help you because I'll be busy supervising the guys in the shop building the robot. *I* think you can do it, Ceira Gisselbeck. You're quite capable on your own."

My thoughts whirled. I'd done most of the work on the table display and oral presentation for the last couple of years and I'd helped since eighth grade as part of the gifted ed program at Conway. As a high school junior, I was now a leader in the club and I liked a challenge, but—*Gibby complimented me to my face.* That realization shocked me more than anything.

As a teacher, he would prod, push, badger and all of a sudden you'd understand the problem and understand him as a great teacher, but he rarely made a direct verbal compliment. I flashed to

Paul and Phil and the Robotics Club people, my good friends. They couldn't take a robot to competition without the mock corporation components! I couldn't let my friends down.

"Sure. I can do it. Bring it on."

"Good. I'm sure you can, too," Mr. Gibson said and smiled. "And you'll have lots of help. Looks like we'll have thirty or more in Robotics this year. With you as the club VP for the corporation component," he paused, "we'll be just fine. Good girl. See you at the club meeting."

He walked away and I felt totally alone even though the bell rang and the halls filled. Friends and freshmen would have jostled me and must have shouted the usual greetings or dumb jokes, but I have absolutely no recall of what any of them said that first day of classes.

Good girl? What- am I a pet?

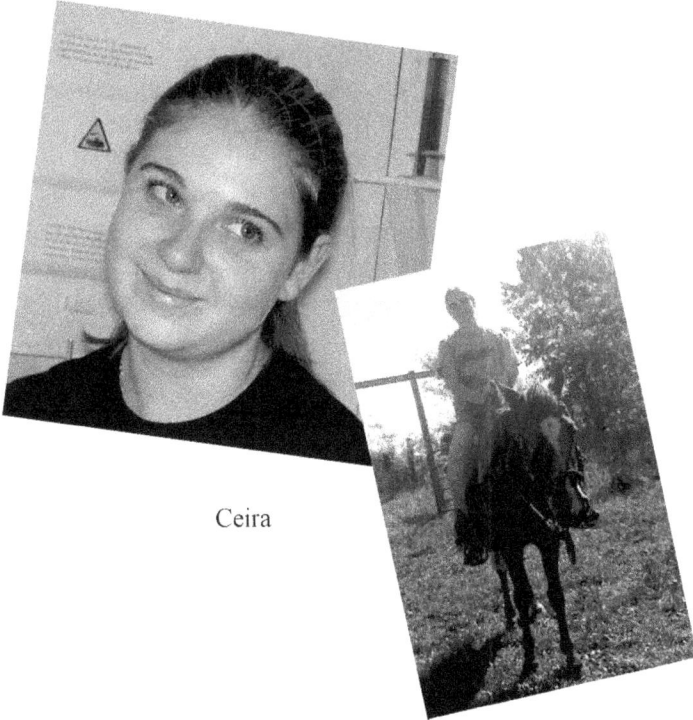

Ceira

THREE

+

CEIRA

I was one of the first in the Robotics Club organizational meeting, anxious to get the year going. We met in Mr. Gibson's room. A group of students had asked him to sponsor the club after it had some problems and went dormant for a couple of years.

Gibby came to CHS to teach math after ten years or so as a building contractor. He also used to be a stage director or actor or some such and even did some modeling. He used his drama skills in everything—including his math teaching. He would laugh one minute and yell the next. People who'd hated math started having fun in class. But in Robotics Club, he went to new levels of drama. I know that sometimes he yelled just to get us fired up.

Mr. Coyle, Club co-sponsor and a part-time Science teacher who'd been in Conway forever, sat in the back row of chairs. He was pretty stoical. Hard to read but a good teacher and a nice person. He nodded "hello" at me and did his half-smile thing. I waved back. A bunch of people came in right after me and sat near Mr. Coyle. Most of them hadn't been in Robotics Club the year before.

Paul and Phil burst into the room, grinning. In some ways, we three were closer than family. Paul could be intense but his great smile cut across his face and it lit up like a 100-watt bulb. When angry—usually because someone wanted to do something really dumb with the robot—he got real quiet. He was thinking of how to politely say the suggestion was totally ignorant. When totally frustrated, he'd pump his fists up and down in the air and go "UNNNNN!" Then he'd take a deep breath and explain the problem.

Phil was fair-skinned, light brown hair, hazel eyes and he slouched his shoulders. When excited or intense, his cheeks got bright pink. He hated that, but the girls thought it was cute.

They were both thin with wiry builds, and just a little bit taller than me. I'm 5'8." Not all the guys were taller than me and some didn't like that. Paul and Phil didn't seem to notice.

My hair is honey blond and I have tri-colored eyes with the inner iris ring gold, and that starbursts into blue with an outer rim of grey. I don't wear makeup. A woman shouldn't hide her inner beauty. Some people think I'm full of myself, but I'm not. Self-confident, yes. Prissy, no.

Truth be told, I had a mild crush on Phil for a while. So did Sara Carnes. No—Sara had a *huge* crush on Phil. I once tried to get him to go out with her but he wouldn't. I called him a coward and he said, "You got that right."

Most of us in robotics didn't do sports, unless you consider marching band or for me, riding my horse was sport enough.

Mr. Gibson loped into the room. His face glowed. He loved Robotics Club in his unique alpha-male, drama-guy, math-nerd way. I heard him say, "Hey Paul…" then his voice lowered while he talked to our CHS robotics genius. Paul grinned and nodded.

More people crammed into the room.

Too many people. A whole ton of people were there just to goof off. That wouldn't last long when they heard all that was involved. The BEST organization had a ton of technical specifications for the robot, the mock corporation, and the competition.

BEST stands for Boosting Energy, Science, and Technology. It's an international organization that started several years ago with engineers at Auburn University and has grown each year as more and more schools get into robotics."

In a small school like ours, you knew everyone. You knew who did well in which classes, who had to work after school *and* would never be around to help with robotics, who dated who from eighth grade on and was only in the club to hunt up a date. You also knew where everyone planned to go after high school graduation and if that was college, those people knew the importance of Robotics Club on their resume.

Mr. Gibson came over to me said, "I'll get this meeting going then turn it over to you, Paul and Phil. Just tell them what's involved in the different club areas and that it's hard work. Briefly go over the job titles then pass around these signup sheets."

He went back to Paul and Phil and left me standing by myself until the two Ashleys, Shyanne Witt, and Sara Carnes came in and stood next to me. Sara smiled toward Phil but he'd turned to Paul soon as she came in the room.

"Hey girl," Shyanne said. "Good to see you." We leaned against the wall toward the front of the room and waited for Mr. Gibson to start the meeting.

"Did you hear that Ms. Day won't be able to help us this year?" I asked Shy.

"Yeah. Bummer," she replied."

Gibby stepped to the front and everyone got quiet except some girl was popping her gum like no one dared to do during class. I frowned at her and looked around the room.

As Gibby went into Mr. Gibson mode and explained about BEST Robotics, some eyes glazed over. When he said there were 10,000 high school and middle school students at several different Robotics Kickoff Days around the USA that year, several people responded with "wow," "awesome," and "whatever" mumbles. The gum popping bitch stopped with her mouth open.

He continued, "Because BEST gives a huge tub of raw materials for building the robot, it's theoretically an even competition for poor as well as wealthy schools. Most of our competition will be wealthy private schools with tons of resources. That's not us. You know that. We'll have to raise most of our own money for trips to competitions. Our school just doesn't have much money for extra-curricular activities and I've been told the state has cut back even more this year."

Groans and mumbles echoed around the room, "Figures."

"Same ole crap."

"Tell me something I didn't know."

"Yeah yeah, we're used to scrounging."

When Mr.Gibson told the group about going to Northark College for Kickoff Day, getting a huge tub of raw materials and a big notebook full of rules for designing and building the robot, playing the game, *and* the corporation component, people's faces changed. Some smiled—the best math and science students. Others looked like he'd thrown cold water in their faces.

He added that we also had to design our own graphic logo for t-shirts, build a table display, a web site, a spirit team, and develop a mock corporation complete with mission statement and officers. The corporate officers would stand in front of a panel of judges and explain our corporation from the get-go to the final ready-to-rumble robot. The judges were engineers and Northark College professors. Robots would have a live game competition. There'd be tryouts for driver soon as our robot was built.

He motioned toward me and said, "Ceira, explain more about the corporation, please."

I looked at the slackers and steeled myself. "See these people standing next to me? We're serious about robotics. Yes, it's fun but it's also a lot of hard work. You have to stay after school every day and work. Weekends, too. So if you're not serious, if you think this is goof off time, it isn't. If you aren't passionate about this club and willing to work hard *after school*, then just get up and walk out the door!"

A few people got up and left.

Good.

Ashley Fields' brother, Broghan, later told me he thought I was a bitch at that meeting. "Why?" I'd asked.

"Because you stood there with no smile on your face and told people that if they weren't prepared to work hard to get the hell outta there and your eyes flashed daggers when you said it. If you remember, several people didn't come to the next meeting."

"That was my intent. That chick with the gum made me crazy and so did the guys who showed up just to join every possible club they could. We didn't need people farting around and getting in the way of those of us who actually cared about Robotics Club. And I didn't say hell."

"You did with your eyes," Broghan said.

That was only Broghan's second year in Robotics so I didn't know him very well but his intensity intrigued me. I'd seen too many people join the club as a lark, thinking it was just going to be a live game thing. They must've thought the robot parts came ready to put together like a model airplane. Broghan knew better.

Students ran everything in Robotics Club. We had one president but a ton of vice-presidents for all the different components. For overall club leadership that year, Phil was club president. Paul, our genius robot builder, was vice-president for the robot. They'd earned their jobs. Paul was born a nerd. He loved math, science, mechanics, and didn't mind all the after-school work. He'd helped build three prior robots and was more than ready to be the key designer. Phil was good with math and science, too, and had helped build several robots.

At the end of the meeting, I felt pretty good. We had a great core group and several eager and serious freshmen that would be good apprentices. We also had a few raw newbies from junior high. Made me feel old.

Mr. Gibson announced club dues for the year were twelve dollars. "If anyone has trouble coming up with that, see Mr. Coyle later," he said.

Our school district was officially categorized as rural underprivileged. Poor, but we didn't say it that way in the documents. I knew that sixty per cent of our students were on free or reduced lunches. That was part of the demographic data we'd dug up for the corporation papers the year before.

Most of us ate breakfast at school, not just the low-income people. Some of my friends didn't get much to eat except at school. But we were also unique because ninety-five per cent of our student body graduated from high school.

When three seniors swaggered into that first meeting, I almost lost it. Gibby's face echoed mine. Paul and Phil looked like they wanted to bellow.

What were those guys doing here? Grant Rumfelt, the jock basketball forward, super shy nice guy Lloyd Oberbeck, and the totally wild Shane Sell elbowed in. Shane was such a pain-in-the-the-everything. Grant hated me. Lloyd was so shy he wouldn't look me in the eye. Shane had been in the club last year and was smart enough with math, an excellent gamer—but he could be such an in-your-face annoyance. Teachers hated him. The lunchroom help hated him. I got along with him okay, but he'd butt heads with Paul. *Why bring his two best friends to Robotics in their senior year?* When I heard them whisper to someone that they'd joined every club they could for their final year at CHS, I was furious. *How dare they? We didn't need them.*

Yeah, right.

I was naïve about a whole lot of things that first meeting. Had no idea we faced a year that would be the emotional and physical equivalent of a Six Flags roller coaster. Didn't plan to fall in love, get my feelings stomped on by Gibby, be dissed by the principal, go to my first prom, and learn a ton about leadership. At that first club meeting, I thought I was ready for the year. *Ha!*

Conway old city limits

Shyanne & Ceira

FOUR

+

PAUL

Robotics Club was a natural thing for me because ever since I could remember, I'd loved to tinker. I'd fix appliances and farm equipment. Sometimes I'd put together odds and ends of scrap metal, bolts, wheels, boards, whatever Dad would let me have, and I'd make new things. Figuring how things worked was easy. But when I had to give up cross-country track for robotics, that was hard.

I hated that there wasn't enough time to do all the great things you got with high school, even one small as ours. Worse, *the principal* didn't understand. He used to be the cross-country coach and he told me to stay in track. "For college, you need to be well-rounded and do athletic stuff as well as academically challenging stuff," he said. He didn't get the importance of robotics for me, for my future.

I gave up track my junior year and focused on robotics. With Mr. Gibson as sponsor, we finally had a great thing going with Robotics Club and I couldn't miss out on that. I'd worked my way up to being the lead designer and builder for the robot.

My parents supported my decision to go against the principal's advice. So did Phillip, my best friend. He was super good with wiring, creative, knew how to repair tons of equipment and could drive anything you'd find in town or on a modern farm. He'd build the robot's wheels while I worked on the overall design.

Philip grew up on one of the few remaining family farms in our region that actually made a decent living. His dad, grandpa, and great-grandpa farmed about 900 acres together, and they milked a ton of cows morning and night in a modern facility.

Phillip had a lot of farm chores, especially in summer. On a working farm, you don't hire a handyman, you fix equipment on the spot. Phillip grew up with tools and a ton of equipment large and small. But his parents knew that an extra-curricular thing like robotics was important so when we got into building the robot after school, he got to miss some farm chores. He'd go home right after school and help milk, then drive the family SUV back to school for robotics.

I, on the other hand, grew up on a 20-acre hobby farm and my parents worked in town. I come from a long line of tinkerers and fixers on my dad's side. They're engineers but sometimes without the formal degree.

My parents supported me in school *and* in Robotics Club. They even sent me to a couple of summer engineering camps at Missouri University of Science & Technology, known locally as S & T, or Rolla, which is the city where it's located.

So Phillip and I both grew up building and repairing things like mowers, farm equipment, household appliances, plumbing, wiring, building sheds and barns. We were both good in math and science classes, too, especially since Mr. Gibson started teaching at CHS. He challenged his students and sometimes made you work harder than you wanted, but we actually loved that, even if we griped about it.

While we did our robotics thing after school, Gibson sometimes called us Doofus and Dingus. He'd use the nicknames interchangeably so we never figured out which one of us was which. Didn't matter. He supported us in Robotics and that meant everything. He encouraged, challenged, and most of all, believed in us.

We called him Mr. Gibson in class, but in Robotics we called him alternately Gibby, Gibson, and The Old Man. Never had so many nicknames for a teacher.

It was great that out of a high school with less than 300 students, we had thirty in Robotics Club. Some of the girls in the club were there for the art work, like the t-shirt design.

Then there was Ceira, who was one-of-a-kind. She grew up helping her dad with his construction business. She could operate more equipment that anyone I knew except maybe Phillip. She drove a tractor, backhoe, a 'dozer and other big equipment. She said the one thing she didn't master was her dad's dump truck. It had thirteen gears that weren't clearly marked and they didn't make sense. But I wouldn't be surprised if one of these days she drove that thing, too.

Ceira loved a challenge like no one else except maybe Phillip and me, and she had the smarts do get the job done. She took all the advanced math and science classes available at CHS. She was also pretty, which made it funny to some people that she'd drive big equipment or be in robotics. They wondered why she wasn't a cheerleader. She liked to break stereotypes and no way would she be a cheerleader. She'd probably rip someone's head off that suggested she be a cheerleader or the band majorette. She was way more into academics than the girly stuff.

Lots of people in Robotics Club liked the glamour of driving a robot—using the controls to make it move and do things—but they just didn't have the mathematics skills to build one. It was more than mathematics, actually. You had to be able to see in your head what the robot had to look like—to envision parts needed to get the job done. You had to figure out what gears you'd need in it, and you had to make those gears as well as the robot body.

Each year it was a different theme for robotics competition, so each robot had to be built specifically for that year's competition. That was the fun part about the BEST Kickoff Day where we'd see the new game field and envision the new robot. That was incredibly exciting.

Before Kickoff Day, BEST always sent out a teaser video for the new theme. They'd put the video on their web site to tantalize you, to make you guess what the theme would be.

The newest teaser video started totally black, and then a few chemicals came on the screen fast, and then more fast flying equations zooming around then off the screen. It was only a twenty second video but it got Phil and my brain engines going—and Gibby's, too. His engines would go faster than anyone's. From the chemical equations and fast movement, some sort of fuel was clearly involved so I speculated the robot would be a race car.

At the first club meeting that year, Gibby introduced Phillip and me as the robot engineers. When we elected officers it was pretty much a done deal that either Phillip or I would be President and the other would be Vice-President for the robot because we were the only two who had enough engineering skills to build the robot. It *was* Robotics Club, after all.

It was also a done deal that Ceira would be the corporation V.P. Phillip and I felt bad that there wasn't a teacher to help her, but she could do it on her own because she'd worked as apprentice then lead on the table display during her years in the club. She was good with English as well as mathematics and chemistry. I didn't much like English.

The first club meeting was the first Wednesday of the school year. Kickoff Day was the following Saturday at Northark College in Harrison, Arkansas. Man, I was anxious and excited.

Working in the hay field

FIVE

+

CEIRA

Kickoff Day was nerdvana, a super exciting event the second Saturday in September. Around the U.S., some 10,000 students went to BEST Kickoff Days somewhere at a college with an engineering department.

The first competition was called the Hub. It was almost always the same place as Kickoff. A couple months later, there would be Regionals for the winners from all the hubs. Our Kickoff and Hub were at Northark College in Harrison, Arkansas. You had to have a robot for competition at the Hub, and you had to have other components like the mock corporation, table display, spirit team, and a web site.

We boarded a school bus at 6 a.m. that Saturday and headed south for our Kickoff day. Laclede County R-I is the official name of our district for our paperwork. But at the event, they'd just refer to us as "Conway."

The mock corporation component required documentation so I had a digital still-shot camera, and a videocam to record the Hub events. Already on the web site and in our notebook, we had photos and descriptions of our robot's design, the construction process, some of our fundraisers, our corporation's mission statement, officers' names, and the school district's demographics. We figured to qualify for Regionals, so we'd add Hub competition photos to our documents.

On the three hour bus ride to Harrison, we sang songs. Someone started with ninety-nine bottles of pop on the wall until Gibby yelled "beer." After that everyone yelled out the whole song. He could be as crazy as any of the high school kids—part of why we got along with him so well. Then someone started the kiddy "Wheels on the Bus" song, then "Bohemian Rhapsody." Only the girls knew all the words to "Rhapsody" but the boys joined in when they could. Even Paul belted out "Gallileo, Gallileo."

After fifty miles of I-44 to Springfield, MO we turned south on Highway 65. Thirty miles passed Branson, we hit the Northark College campus in Harrison, Arkansas. Three hours of bus ride went pretty fast when surrounded by your best friends.

Super-charged students clambered off busses, blinked in the sunlight, climbed up *long* steps, and charged into the Northark field house. Students and teachers from a dozen schools in Missouri and Arkansas swarmed the field house floor and bleachers. Everyone was anxious to see the mockup of the new playing field and see this year's theme.

As my eyes adjusted from the bright sunlight to the field house lights, a big banner hanging from the upper tier of the field house demanded attention: HIGH OCTANE. The new theme.

I focused my camera. *Click, click.*

High energy generated by students plus teachers and a sprinkling of parents created a force field. Excitement fueled the people but electronic brains would fuel the yet-to-be-built robots.

Mr. Gibson and other teachers lined up to register their schools and pick up the rules for the High Octane year.

Click click.

Mr. Gibson chatted with other teachers and the event organizers, Northark College engineers. He also talked community business leaders who were there as event sponsors. He'd made it a point in previous years to get acquainted with other teachers, not only to be polite, but to pick their brains about robotics. He also *always* talked with the Northark engineers who built the sample field and who knew the High Octane game rules.

At these events, Mr. Gibson was totally professional with none of the kidding around he did in our after-school activities.

The new competition field was laid out on a large piece of industrial carpet on the field house floor. The field was where the robots would compete in six weeks. I saw some tennis balls and racquet balls rolling around the floor and a plastic inflated globe hanging over one side.

I took pictures from various angles and exchanged greetings with friends from other schools we knew from prior competitions. A few parents walked through the playing field but most left the serious field perusal to robotics students and teachers.

The two senior interlopers followed Shane, who lost no time looking for cute girls.

Grant scowled each time we made eye contact. I guess I *may* have scowled first. I walked past him, doing my camera thing to document the activities. Lloyd walked with Grant but he always smiled at me and I smiled back.

Mr. Coyle walked around the field, hands behind his back, eyes focused on the setup.

Paul's face showed intensity like no one else examining the new playing field. His focus was amazing. Phil walked with him. They stopped at certain places, pointed, talked, then moved on. They covered the field and then revisited each area.

Click click.

I took my camera to where large six foot folding tables filled one end of the field house, behind the registration table. Conway was # 185.

From past competitions, I knew that the BEST organization gave eighty-five percent of the total points to the corporation elements of the competition for what they called the BEST Award. My part. But damnitall, the guys driving the robot always got the most attention!

I watched Paul's intensity deepen with each step around that game field. *Click.* Mr. Gibson joined Paul and Phil. *Click* They talked. He nodded his head, spoke, pointed to the hanging globe, then to the balls on the floor.

Click click.

Other Conway students, including Broghan *my heart fluttered* joined the huddle that had formed around Paul. The group broke apart and some walked around the field and stopped at the corner of the game field, did more pointing and talking.

Other teams walked, grouped, talked, and looked. Most wore their team t-shirts from previous competitions.

I switched to video for a few minutes and scanned the field mockup, wondering what chemical components the cans, balls, and globe represented for High Octane.

The field house PA system interrupted with, "Everyone please take your seats and we'll get this event rolling."

Rock music blared as the announcer grinned and motioned to the crowd. Students and teachers scattered, and then re-formed groups in the stands.

Paul's parents had saved seats for Conway. Mrs. Coryell's camcorder whirred. She smiled. Paul's sister Rachel, a 7th grader, waved. The little brother, a preschooler, jumped up and down.

Mrs. Gibson looked up from the book she'd been reading and smiled as Mr. Gibson sat down next to her.

The music abruptly stopped and the announcer took the microphone. "I want to tell you about this year's event." He paused a couple of moments while the last people climbed to their seats. "But first, on behalf of Northark College," his voice picked up intensity, "I want to personally welcome Ev-VVV-ER-Y ONE to Kickoff DAYYyyy!"

Applause and cheers erupted from the stands. The stadium vibrated with foot stomps and ear-splitting rock music. I turned off the video and cheered too.

When the crowd noise subsided, the announcer gave simplified play-by-play action for the High Octane competition while some of his fellow engineer-professor colleagues demonstrated challenges for the robot on the playing field.

Tomato paste cans were dispensed from a central carousel. Tennis balls dropped from PVC pipe contraptions and rolled around the floor along with racquet balls.

That's a challenging set of objects for the robot to collect. Wonder how Paul will design this year's robot. What will I make for our team's table display? Maybe a mock chemical lab.

I took a picture of the announcer. He added details. Paul's focus riveted on the narrative. I think he absorbed every word. Gibby's attention also riveted on the speaker.

For this year's competition, the robot collected items that represented chemicals. The right combination of chemicals would make an energy-efficient, high octane fuel. Each team would begin with their robot in the corner of their quadrant next to a mock fuel refinery. The scoreboard buzzer would ring the start and end of each round. The scoreboard *BZZZZLATTT!!* punctuated his words.

The playing field would have four robots going at once. Each three-minute round would focus on the robot, controlled by a driver with a wireless hand-held controller and on the side of the field near the back corner would be a team member called the spotter. Items collected by the robot had to be deposited into the fuel refinery. The spotter gave hand signals to the robot driver and was the only team member who could touch the items.

Racquet balls represented H_2O, water. Plastic globes represented CO_2, carbon dioxide. Tomato paste cans represented energy. Tennis balls represented the catalyst. Each team's score would be determined by the combination of items collected to make a chemically correct, energy efficient fuel—High Octane.

The robot driver must think, plan ahead, watch the competition and the team spotter, dodge cans and balls on the playing field, pull down a plastic globe and deposit items in the refinery while maneuvering the robot with a hand-held controller. Drivers would have to use chemistry and mathematics skills on the go along with hand-eye gaming skills.

Competition would begin with elimination rounds, then semi-final rounds with eight robots, then the final rounds with only four robots.

The announcer reminded the crowd that four robot teams would simultaneously compete on the four-quadrant field in each round. His face morphed into a grin and he said that on the game field, robots could block other robots. But any robot that got into a competitor's playing field and was tagged by the other robot had to freeze for thirty seconds. In a three-minute round, that freeze could make or break the winner.

On opposite sides of the Conway group, Shane grinned. Paul frowned.

The top eight teams at the end of elimination round would advance to semi-final rounds. A benzene tanker would be added then, and its proper use could considerably enhance scores.

The top four scoring robot teams from semifinals would play in the final four rounds of competition. At the end of the day's competitions, only one robot would be the Hub competition's top dog. Trophies would go to first, second, third place robots. Award plaques would go to the best designed robot and the best spirit team.

The biggest trophy, the BEST award, would go to the team with the most outstanding corporation component.

The afternoon before the robot competition, judges would score the components of the mock corporations and table display, my gig. Of course, I'd have helpers and other people would design the t-shirt, do the web site, and compile the notebook. My helpers and I would work on all that while Paul and Phil built the robot with some apprentices trying to not get in the way. Broghan would help with the robot. He was smart enough to know when to speak up and when to listen.

The top three performing robots and corporation teams from the Hub would go on to regional competitions towards the end of the semester. The announcer smiled, and said that schools from eight states would compete at our area regionals held at Arkansas State University in Ft. Smith.

My challenge just got bigger.

Gibby & Ceira listening to announcer at Northark

SIX

+

PAUL

That flying molecule teaser video had haunted me ever since Gibby showed it to us at club organization day. We talked about it on the bus all the way to Kickoff Day except for the time singing stupid songs. Gibby sang too!

We finally got to the campus and into the field house. I saw the banner at once, HIGH OCTANE. My thoughts raced. *So I was right—some sort of car race.*

Gibby went to register our school. Phillip and I headed for the mockup of the new game. As soon as I saw the field, I knew it was not a race track. There were soup cans and tennis balls rolling around, and a big globe hanging over the side. *Something to do with space? BEST already had Rocket to Mars as a theme one year so it had to be a new angle, but what?*

Phillip and I walked around the field a few times, trying to figure out the point of this new game. After a couple of rounds, we went to check our raw supplies to see if we'd get a clue there.

Forty-gallon plastic bins on six-foot tables contained hardware such as screws, nuts and bolts, an irrigation valve cover, piano wire, an aluminum paint grid, a bicycle inner tube, the micro-energy chain system and a BRAIN. BRAIN stands for the BEST Robotics Advanced Instruction Node Programmable Platform. The computer inside the robot. On and under big tables were the materials for each school's robot. Most were same as we'd had in

28

the years before, like pieces of plywood, aluminum, and PVC pipe.

Although the brain and motors had to be returned at the end of the competition, other parts were consumables—items that would be sawed, drilled, milled and made into working robots. We wouldn't have to use every single piece in the kit, but couldn't use anything that *wasn't* supplied by BEST. It all looked like other years' raw materials.

I'd been at the kickoff before and knew the judges used a checklist of items when they examined each robot at competition check in. They counted nuts, bolts, screws, rubber bands and even examined tape. If a part broke during practice or competition, it could be replaced with identical components purchased at the schools' expense. Key word: identical.

Back to the field.

Out of the corner of my eye, I saw Ceira taking pictures but I didn't stop to talk because my head was spinning with robot ideas. Phillip and I kept walking around the High Octane field. Gibby joined us and we talked about all the different things the new robot would have to gather. There was more variety in sizes and shapes of objects to gather than we'd ever seen before. This game was more complicated than anything I'd expected. I couldn't envision how one robot could pick up all those objects.

My brain engine whizzed. I had a basic idea of what the new robot was going to look like. It would need an arm with some sort of adjustable holder that would grab those different sized cans, balls and the globes. This robot would be way more difficult than any other I'd built or helped with.

Phillip, Gibby, and I were the last people on the playing field when the announcer started his welcome stuff. We hustled to the stands.

We knew the basics, like the first competition would take place in only six weeks. Six weeks to design and build the robot, develop the corporation, the table display, the web site, the t-shirt, the spirit team and the PowerPoint presentation. Six weeks of almost non-stop after school robot work.

The announcer gave a play-by-play about the High Octane game while engineers walked around the playing field, dispensing the different items.

Turned out, High Octane wasn't anything about a race except to gather more points faster than other teams. It was an environmental theme. The goal was to gather things that represented chemicals to make a high octane energy-efficient fuel. Isooctane.

When the floor demonstration ended, the announcer thanked this one and that one for doing this or that, and the students, including me, only half-listened. Someone in the business community donated something. A group hosted a dinner for teachers involved in robotics. Polite applause punctuated at appropriate moments.

Students looked toward the concession stand. Some were texting. The announcer called someone to the mic for a special tribute award. Blah blah.

After lunch, there'd be small group meetings for the robot builders and other meetings for the corporation groups. The announcer wished each school good luck, then stepped aside.

Rock music blared as students and teachers thundered out of the stands and collected their raw materials or jammed the food stand. We took our supplies to the bus then returned to the field house for the small group meetings.

On the bus ride home, we talked about the new game. I got irritated at some of the talk. Some people speculated how they would build a robot that could pick up the crazy variety of objects in the new game. Sometimes I tried to explain why the idea wouldn't work and other times I just ignored them.

We got back to CHS late afternoon, stored the raw materials in The Old Man's classroom, and a few of us stayed to brainstorm.

Over the weekend, I sketched some designs for the new robot.

Mr. Gibson always built a mockup of the year's playing field at his house. BEST didn't provide the raw parts for practice fields. He used some of his own materials left from when he worked as a building contractor, and some items donated through his contractor connections at Lowe's and lumber yards. The field Gibby built would have all the basics of the High Octane official field. He'd haul it to school where we'd lay it out on a square of industrial carpet. The carpet helped control the balls and cans on the field, and it kept the robot's wheels from spinning.

At the end of the school day, we'd move chairs to one side of Gibby's room, spread out the carpet, then set up the field for practice. But practice couldn't start until the robot was built.

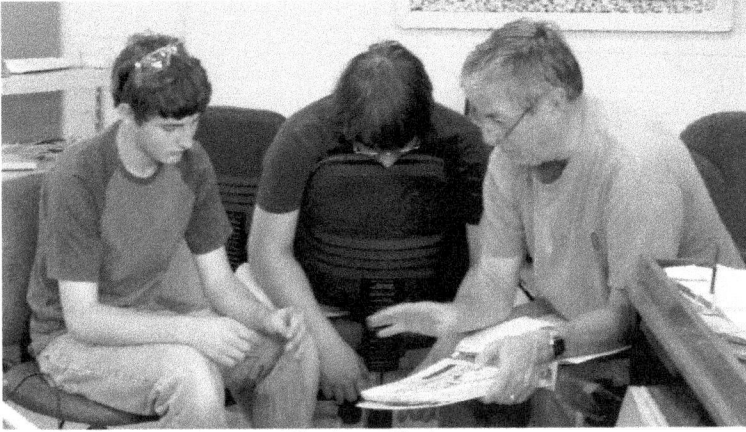

Design talk time.
Paul, Phillip & Mr. Gibson

SEVEN

+

CEIRA

On Monday after Kickoff Day, only the serious robotics people gathered in Mr. Gibson's classroom for club meeting time, the last twenty minutes of the school day. Robotics Club was run by the students but not in typical committee structure like other clubs. The president was sort of a figure head role, a tribute to who'd worked up the ranks, mostly in building and competing with robots for four or more years. Instead of committees, we had vice-presidents in charge of the robot, the t-shirt, the web site, the notebook, the table display, and fund-raising.

Even though Gibby said I was in charge of everything Ms. Day used to oversee in the corporation, he wasn't shy about telling me when he thought something should be different. But I stood my ground. I could be a drama queen to match his drama king. If he yelled, so did I. In the long run, I think he respected my strength.

To get the different parts of the corporation stuff going, we V-Ps started with a lot of "what if" discussions. We were used to the fast pace of the first weeks where we had to pull it all together. The corporation pieces and the robot would be modified after the Hub event unless we didn't get to go to regionals, but we usually did.

We buckled down and got our stuff organized.

For the table display, I stuck with my original idea of a chemistry lab. I'd beg and borrow from the school's lab as much as possible. Building the actual ten foot square booth with sides and a top would come mostly from building supplies I could scrounge because the BEST organization didn't provide supplies for the corporation. Fortunately, my dad owned a lumber yard and he was good about supporting my school activities. He donated lumber, canvas, poster board, nails, and paint.

In addition to organizing the table display, I took photos of the robot design process where Paul, Phil and the wannabe robot builders tossed around ideas. After their idea sessions, Paul would sketch out designs. Broghan put the designs into the computer using a special BEST program for our web page.

Sometimes Mr. Gibson had to take a turn monitoring detention in his room while we robotics rowdies did our thing. It got a little crowded in there and the detainees got interested in our stuff so Gibby would refocus them on homework. He was good at explaining math to those kids, like telling one guy to think of knocking off a head while multiplying fractions. I smiled when I heard him say that.

Bear Time—what the admin called club meeting time during the school day—was too short for all the things we had to do in those jam-packed six weeks leading up to the Hub. Those who had cars, or could beg a ride home stayed to do more work after school. Some of us stayed until 8 p.m. or later and left only because homework or chores waited.

I had a triple hand-me-down ancient black Honda and I always drove some people home, usually Ashley Fields and her hunky cute brother Broghan. Those two had to take turns at after-school robotics while the other one baby sat their little brother and two sisters. Ashley knew I liked Broghan and at first, talked nice about me to him. He was beginning to make eye contact and smile at me and my heart fluttered each time.

After school, we moved our groups to where we had the necessary equipment or space. The web site group moved to the computer classroom adjacent to the shop. The robot boys moved into the shop. The t-shirt design group went to the art room and others just disappeared.

One guy who shall remain nameless showed up only when Mr. Coyle brought food. Nameless had a nose better than a shark. It was unreal.

My speech had scared off the younger goof offs but not the

seniors pals of Shane. He, Grant, and Lloyd, jumped into the action as if club veterans. Sometimes that helped, other times it was near-disaster. I watched some robot idea sessions when Shane threw out ideas and argued for and against most everyone. He never knew when to shut up. Paul and Phil stood their ground with faces red, eyes flashing, and their voices all got loud. Phil's cheeks got redder with each confrontation.

Grant missed a lot of the robot design time because he was a hotshot varsity basketball player and had to practice first, then do robotics. He drove a new car, which bugged the crap outta me. Yeah, I was jealous. Why should he have a new car when I had a family hand-me-down junker?

Mr. Coyle brought food to late after-school sessions. His cooking repertoire was limited but still we appreciated whatever he brought. One week he brought a pot of chili several nights in a row. We joked that he just added water to make it last from about day two through day five. Even when we got really tired of it, we ate. It was food!

Some people brought their own chips and dip. They'd write their names on the dip containers and shout that they'd contaminated the carton with double-dipping. Theoretically, this kept others out of their stuff. Didn't stop Shane or Gibby. Gibby double and triple dipped a hot dog into the chili pot!

Phil issued threats of bodily harm if his chip dip was eaten. *Ha!* That had the effect of waving a red flag in front of a bull for some of the guys. Most of the girls wouldn't double-dip, though. I mean, really!

Bob Gibson

EIGHT

+

PAUL

Robot design sessions made me crazy when people started arguing. Mostly I stood my ground, but other times I'd see something in a new way, a good way. The tasks demanded for the robot in the High Octane competition were huge. I could envision but not as easily explain what I could see the robot needed.

I wasn't worried about the chemistry, that was easy. It was trying to figure out how the robot would collect all those different shapes that gave me fits.

At first, a bunch of people who didn't have the background came to the design meetings and blurted out ideas. Some of the suggestions were dumb but I tried to be nice, especially to the younger kids who were there to learn. But I swear that Shane sometimes threw out ideas just to be argumentative.

There was yelling and people interrupting each other and some simultaneous talking and yelling so fast you couldn't always tell who said what.

Gibby: "We've got to have an arm with some sort of hook that can pick up the slick racket balls, and the fuzzy tennis balls, plus the soup cans and get all that to the holders in the refinery."

Phillip: "Don't forget the inflatable globes. How're you going to pick them up with just one arm?"

Someone: "We'll need two arms, then."

Me: "Not if we have one arm with an expandable holder that

can adjust to the size of each object."

Phillip: "The globes will be hanging. And they are much bigger than the other items. The robot has to reach up and take hold, then pull them down."

Me: "Yeah."

Someone: "Those plastic globes are huge compared to tennis balls. And slick."

Phillip: "How will the globes attach to the lines?"

Gibby: "Dunno yet. I'll find out."

A freshman: "And it has to pick up tomato paste cans and racquet balls."

Me: "Okay."

Someone: "We'll need two arms."

Me: "Like I said. With the holders on the end of the robot's arm. We could make some sort of grippers that could open wide for the globe and close up small enough for the other stuff."

Nameless: "Huh? Guess I'll have to see that."

Me: "You will."

Gibby: "The arm will need to extend and retract, too. It will have to reach out for stuff and then fold up to fit the box."

Some new guy: "The box?"

Phillip: "Yeah. The box that the judges fit the robots into. The cube that measures two feet by two by two."

Newbie: "Oh. That box."

Shane: "Yeah, dummy THAT box."

Newbie "Don't dummy me you moron."

Shane: "Yeah, yeah."

Phillip: "Gibby, What do you think about the arm?"

Gibby: "Paul, do you have the robot arm figured out?"

Me: "I'm working on it."

The Old Man stopped everything when he asked, "What are we going to call this year's robot? It's the turn for a boy name."

Phillip: "Remember we alternate gender each year. No sissy names this time, like Ling-Ling."

Several voices: "Yeah!"

ME: "Fred"

Gibby: "Huh? Where'd you get that?"

Me: "Fuel. Replenishing. Energy. Efficient. Droid. FRED."

Phillip: "Awesome."

Gibby: "That's good Paul, really good."

I smiled.

The Old Man expanded on the Northark announcer's overview. He'd studied the rules of the High Octane game, and I knew that if anything in the rulebook was unclear as we worked in the weeks getting ready for competition, he'd query BEST officials. He'd made up his own three-ring binder with tabs and he knew more of the details than anyone, including some of the judges. He'd have that notebook ready for every design, building and practice session. Even better, he *always* had it handy during competition.

My thoughts focused on the robot design ideas while Gibby went on about the rules.

He'd badger us to learn the rules because we had to apply the rules in the game. "No time to stop the action and consult the rules during game time. Use your heads!"

He was right, but that didn't make it fun. "Besides, he reminded us, "during competition teachers were forbidden to coach from the floor. It would be all student actions. That is, all student territory except for the judges."

Scoring got complicated during competitions. If the team that won the BEST award had a robot that came in fourth at the hub competition, they would still receive an invitation to compete at regionals. Conversely, if a school's robot finished in the top three but the school didn't do well with the BEST component, the robot could be invited to the next level of competition.

Yeah, confusing logic but it meant that smaller schools with decent robots could hold their ground with the big schools who had the resources to make fancy notebooks and table displays. You see, the BEST award encompassed many components used to determine a "wild card" robot team in the finals. Ceira and her helpers had a big job, no kidding about that.

Additional elements judged at competition included each school's spirit and sportsmanship. Spirit teams were much like booster club sections at basketball games. Members wore team t-shirts, had noisemakers, face paint, pompons, and mascots. Sportsmanship meant not only how the teams behaved in the stands, but how the driver-spotter teams behaved on the competition field and in the pit areas.

All I really cared about was the robot.

The BEST award stuff was in the back of my mind. I didn't care a whole lot about how the t-shirt looked as long as it wasn't girly. Phil was good with t-shirt design ideas, but he was too busy helping me with the robot to add much to that group.

The robot design haunted me. This was the biggest challenge I'd ever faced and only Phillip knew how worried I was. I hadn't done anything as complex as this in summer engineering school at S & T. Although Gibby was smart and creative, he wasn't an engineer and so he was limited in his suggestions.

I'd think about it and I'd sketch out more ideas by myself at home, without interruptions from people who talked without thinking.

Paul thinking

NINE

✦

CEIRA

I personally think that BEST should provide the raw materials for the corporation award just like they do for the robot and make everyone use the same stuff like they do with the robot. I mean, some of the wealthy schools could afford really spiffy corporation materials while little schools like Conway could not. Although the BEST organization said the corporation part was the majority of the total score, they only provided supplies for building the robot. But who am I to say the rules aren't fair?

Supply acquisition and fund-raising for the mock corporation rankled me big time with Mr. Gibson. He solicited people he knew for stuff they needed for the robot. He had contacts from during his years as a building contractor.

I had to scrounge for materials. Since my dad donated a whole hunk of the supplies for the table display and corporation presentation, I insisted that Gisselbeck Lumber be one of the sponsors printed on our team t-shirts. Mr. Gibson started to argue over that but I backed him down fast. When I had logic behind my arguments, he'd listen but sometimes things got loud.

Another time I got really angry at *Mister Gibson* was when he wouldn't let me or any other girls use the shop's table saw. I grew up with a construction business! Shyanne grew up in a wood shop, for cryin' out loud! We told him we used saws at home, along with drills, hammers, electric screwdrivers and other power tools. He still said no. I got so angry I yelled and called him a sexist and stomped out of the shop, with Shy right behind me.

We didn't do any more work that day.

We never seriously thought about quitting robotics when we got mad at him because that would mean we'd been sissy wimpy girls. So, when Mr. Gibson, had to monitor for detention in his Math room, and Mr. Coyle was supervising guys in the shop, we ladies got our table display built. We may have had to use subterfuge, but we got the job done. And, we wore our safety glasses and knew how to *not* cut our fingers off!

After I got the table display organized, I started to hang with Paul and Phil in the shop and flirt with Broghan when he didn't have to babysit, and was there. He was one of the smart guys who actually contributed to the design and didn't just blurt out wild ideas.

One day when I went into the shop Broghan was sitting on a heavy table, head down. I sat on the table next to him and bumped him with my elbow.

"Hey, there. Smile" I said.

Silence.

"Ah, c'mon, smile for me."

Silence.

I did it a couple more times, then leaned down so I could make eye contact underneath his long hair. He grinned. I grinned back.

My heart zoomed.

Ceira in CHS shop with Paul and Phil

TEN

✦

PAUL

The design I sketched focused on the robot body and although I knew there would be some sort of arm, I couldn't figure all the intricacies in the initial design phase. Thankfully, Gibson agreed the basic robot body could be almost identical to the one I'd designed the year before. We started building that the day after Kickoff.

In the school's shop Phillip worked on the wheels and I or someone I trusted measured, sawed, drilled, screwed, and glued the robot body into shape. Sometimes we goofed and had to re-measure and saw another piece. Gibson yelled at us when he thought we'd wasted supplies. It's not like we deliberately missed a lick with the saw. He got us extra stuff, but I brought my own table saw and other tools from home, so I put a lot of personal money into the project too.

One time when he was quietly grading papers and we needed to melt the ends of a piece of nylon twine to keep it from raveling, no one had a lighter so we used the blow torch. Phillip thought we were pretty innovative to do that.

Once, someone didn't adjust the saw blade between cutting a piece of plywood and Plexiglas. Oops. Another wasted part. We shoved it into the trash can under some stuff before Gibby noticed.

Other routine actions Phillip and I did during robotics work:

Yelled at Shane after he did something dumb.

Joked with Ceira and Broghan.

Brainstormed robot arm design with Gibson.

Argued with Gibson about the scoop that we decided to put on the underside of Fred to gather up racquet balls, tennis balls, and tomato paste cans.

That scoop idea took a lot of the pressure off me in figuring out the robot's arm. But still, Fred needed a retractable arm to reach up and grab the globes that would hang from fishing line strung overhead, at the outer edge of the playing field.

First wheels.

Paul.

ELEVEN

+

CEIRA

Fund-raising. Gag! I hated begging people for money but we didn't have squat for club funds. The High Octane year found us even poorer than usual. Sometimes we got a little money from the school at the start of the school year, but the state had cut funding to schools so zilch this time. The long standing community priority for anything extra-curricular was sports and next was band. We didn't have parents calling the school board to lobby for funds for Robotics Club.

Our local school district's tax base was small because we had no industry. Public schools were funded mostly from local property taxes. A couple of gas stations, one café, one tea room, and one grocery store didn't generate much tax, not like if we'd had a few factories in our county like Boeing or Ford.

Sixty percent of our students were on free lunches so even in good years, our school district was poor. The school could pay for a bus and a driver to the Hub competition, but we had to buy our own food. If we went down the night before, to avoid leaving Conway at 5 am, we'd have to raise the funds for a motel. The bus was your garden variety big yellow school bus, not a comfy charter with A/C and video players like some of the schools had.

Our first fund-raiser took place at Conway's Community Days, the first weekend after Labor Day. A local farmer donated a side of freshly-butchered beef and we sold raffle tickets for that. People gladly gave five bucks for a chance to win seven-hundred dollars worth of beef. Gibby donated a freezer he was going to put in a yard sale. He's good that way.

Community Days was a big deal. Even though the town is small, there's a lot of spirit and pride. Whoever started calling small towns "sleepy" was just ignorant.

Clubs like ours, and moms with craft stuff, set up booths along one of the streets and sold ice cream, pop, flower arrangements, gizmos, jellies, cakes, pies and burgers. *Mmmm* the smell of sizzling burgers filled the block.

Community Days must be on the circuit for people who make a living by going around various small town fairs and selling whirlygigs, huge wooden American flags, chainsaw carved eagles and other er, crap. One guy sold blow-dart type guns that used miniature marshmallows for ammo. You can imagine the residue of that on asphalt pavement in ninety-five degree Midwestern heat after thirty or forty tweens ran havoc.

Phil and I had to march in the band and we thankfully led the 9 a.m. parade. We didn't want to follow any horses! They didn't have the diaper type things you see on the horses in the Rose Parade.

Phil played percussion and I was the one and only trombone player. Behind the band were a whole lot of little girls in dance tutus, scouts in uniform, at least twenty antique tractors, including one driven by a woman wearing a big purple hat. Next came the Community Days queen and her attendants atop a convertible and then a fire engine with the volunteer fireman tossing candy into the crowd. Last were the show horses clop-clopping and plop-plopping along.

After the parade I checked out the booths and bought some ice cream. I didn't do a whole lot in our booth. I left that fun to those who actually like to hawk raffle tickets on the streets of Conway in blazing September sun—and some did have a great time doing that.

At the end of that blistering hot Saturday, the club had $250 for the checkbook. That paid for about four rooms in the cheap-o motels we stayed in during competitions. But hey, even a cheap-o motel with friends and *without* parents was fun.

Monday at school, fueled by junk food and bad jokes, we again dug into the work of prep for the Hub competition. Sara Carnes did most of the t-shirt design stuff and Ashley Underwood worked on the notebook.

Sara and others drafted designs and tossed, then drafted more. They eventually settled on her design that incorporated the school's bear mascot with a wide open mouth that snarled. The hair on the back of its head flared out to simulate fast movement. The bear went on the oval end of a graphic tanker truck that had the words "HIGH OCTANE" on the side. Above the tanker, block letters spelled out "CONWAY" and below the tanker, "ROBOTICS." It looked good on black t-shirts and polo type shirts. Professional.

The next Saturday, I borrowed my dad's pickup, crammed five people in it, and we went walnut gathering. It's a tradition in our area, and a typical club money-making event. Black Walnut trees grow in everyone's yard and in the edge of wild woodsy areas. The yummy kernels found in stores labeled "Hammon's Black Walnuts" start out a nasty greenish-black casing the size of a baseball. Inside the casing is a golf ball sized hard, black nut. Inside that is the kernel, the yummy part you find in bags at the supermarket.

We wore boots and gloves, walked around yards and fields with five-gallon buckets that we filled with the icky greenish-cased walnuts. Gathering walnuts was a nasty business and if you didn't wear gloves, your hands would be stained black for weeks. You could cover up stained fingernails with super dark nail polish but your knuckles would still look nasty. A couple years earlier, I'd learned that the hard way.

When the buckets were full, we dumped them into the pickup bed. When we had the truck bed mounded too high to add more, we drove to Lebanon or Marshfield to a Hammon's vendor. Their machines shelled the outer casing off the walnuts down to the black nuts and that's what they weighed. But first, a couple of us had a low tech job. We climbed into the truck bed and kicked out the walnuts.

At $9 per pound, the average pickup load brought $70 for our club treasury. We got a couple of loads that day.

I took more pictures of the robot construction process. It was fun to see the progress and coincidentally, Broghan was usually with that group.

Black Walnuts for Hammons

FRED T-shirt

TWELVE

+

PAUL

Gibby came to one after-school meeting with a three-ring binder full of High Octane game rules. It was a little intimidating. We knew this year's robot was going to have to do way more than ever before, but we didn't know how each component dealt with energy. Gibby recapped and summarized. He was really good as translating the professor-engineer jargon into ordinary language.

As in past competitions, we would have robot drivers and game field spotters. But the spotter would be more important than ever before. He would be on the game field in a boxed-in space marked by tape on the floor. The spotter would organize the materials, the faux chemicals the robot brought to him, so he'd have to keep track of all the chemical parts and what equations they represented. He was the only one who could touch the cans, balls, globes with his hands. He'd put the "chemicals" into appropriate receptacles to make High Octane fuel.

To put it another way, the High Octane refinery was a chemical factory. At the end of each round, the judges would sort the chemicals in each refinery to tally each team's scores.

While I worked on the robot body and arm design, Phillip built the wheels. He'd built wheels the year before so we made those first. The parts kit had a bicycle inner tube in it that I exchanged for a used one from home, because why waste a good one? Philip cut two round pieces out of our kit's plywood. Then he cut strips of

47

inner tube and sandwiched those in-between the two wooden wheel layers. He made it so there would be rubber hitting the field's floor instead of wood.

Because unknown shop slackers had damaged the school's saw, almost every day I'd bring my own band saw to school so we could make accurate cuts. We crafted a trapezoidal shaped plywood base for the robot's basic body.

I had a metal lathe in our workshop at home that I used there. The school's lathe wasn't accurate. I'd work on a part at home, bring it back to school, and then install it in the robot. That's what I did to make the small caster wheel that went on the back of the robot's body.

It took about two weeks to get enough basic parts for the robot so we could start testing and fine-tuning. We had a motor on each wheel towards the front for movement of the robot and two smaller motors connected with a shaft I made at home. A string attached to the shaft would raise the robot's arm. Two smaller motors controlled the subassembly's wheels. The subassembly would hold some of the items the robot collected during each round of play.

We installed two servos—small motors—to operate the arm. The third servo from the parts kit would operate a hook that we'd need for grabbing the subassembly. The subassembly was sort of a wagon for Fred to pull after the spotter filled it with the correct combination of chemicals.

We made an underbody scoop to collect the tomato paste cans, tennis balls, and racquet balls so we could drag them to the spotter's corner. The scoop was very close to the floor. The robot could attach to the subassembly and take it to a designated drop off area.

We tested the motors provided in our parts kit at Kickoff Day. They didn't work very well. We traded out at least four. That is, we sent them back to BEST for replacement. Mr. Gibson finally decided we had to buy our own. Those motors had been recycled many times and none of them worked well any more. We didn't want to get caught during competition with a motor malfunction.

Gibby decided to call the manufacturer and get a new motor identical to those provided by BEST. I'd told him I could take apart the latest BEST replacement motor and repair it and he said "yeah yeah" but I found out he hadn't really listened to me.

While I worked on the motor, his back was turned. I had the problem fixed as he was on the phone and I heard him say, "No, we've never taken the motor apart." Then he turned around and—*oops*! He glared at me. Soon as he got off the phone, he yelled, "You made a liar out of me!" He was really mad. But I thought he'd said it was okay for me to work on the motor! And, I'd fixed it while he ordered a new one, which was good, because now we had a backup motor with the one I repaired. But we didn't talk for a day or so until he cooled down.

I don't know if he played on their sympathy for a tiny rural school or what, but he got the new motor just for cost.

Part by part, our High Octane robot took shape. Except that I still didn't have the robotic arm figured out.

At this point, no one knew the High Octane rules very well. Gibby badgered and quizzed us over and over. One time, he ordered all work on the robot to stop until we could quote some of the basic rules. I didn't like it, because I knew we'd remember better if we were quizzed as we actually practiced the game, but he insisted. I knew when to shut up and just go along.

Then we went back to work on robot construction.

Late one evening, when slap-happy tired, we made up a silly song.

Let's build a robot, let's name it Fred
Let's build a robot, let's paint it red.

Let's build a robot, let's make it fast
Let's build a robot that won't come in last.

We build a robot pretty good
We build a robot out of wood.

We build a robot using tape and glue
Not to mention duct tape too.

Let's build a robot better than the best
Let's build a robot to whip the rest.

We'll build a robot better than yours
We'll build a robot to do the chores.
 - by Phillip Foust and Paul Coryell

When Fred, had wheels, a body, a detachable hopper and the new motor in place, it was time to program the brain. No one in the R-I school district or the community, had the skills to do the programming, but Mr. Gibson had enlisted the help of a Missouri State University professor in the past and we figured he'd do the job for us again.

That didn't happen and the whole robotics year almost ended right there.

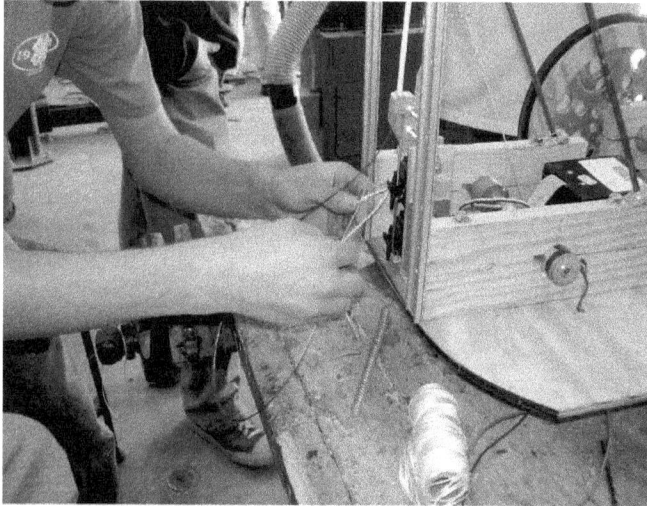

THIRTEEN

✦

CEIRA

We worked our buns off getting the corporation, t-shirts, table display and web site ready, totally unaware the High Octane year almost went bust, but you can't keep a secret long when surrounded by high school students. When we heard about the problems with finding a programmer for Fred's brain and how close we came to not having a functioning robot, I got more than a little upset. How long would he have let us go on without telling? I stomped around some, but realized he wouldn't have let us go on forever. He respected us more than that.

I'd noticed that Gibby had been pulling his hair more than usual, a signal that he was super stressed. For a while I'd thought his grumpiness and hair pulls had to do with management of high school students in building a robot and setting up a mock corporation on a short time frame. That time, a rare thing, I was wrong.

Turned out the MSU professor who'd programmed robots for us before was super busy because of department cuts so he couldn't help. The state's budget cuts didn't just hit public schools. Other professors were equally busy, but he asked some graduate students. None of the grad students had time or interest. It wasn't like they'd get class credit for programming Fred. Gibby had to find someone willing to donate his or her time.

MSU is coincidentally where Lisa Gibson, Mrs. Gibson,

worked as the Assistant Director of International Services so I guess Gibby got the names of some Computer Sciences people from her.

While Gibby hunted for a programmer, Paul and Phil continued work on Fred's robotic arm and the hopper attachment and I kept documenting the action with pictures. Yeah, I flirted with Broghan, who now flirted back. That boy was good at playing hard to get.

Finally, Gibby found Henry Stratmann, an undergrad MSU Computer Sciences major willing to do the programming.

Henry came to CHS a couple of times. He wanted to see the school and meet the Robotics Club people. I think he expected only guys to be in involved. Funny.

Henry was shy around everyone but super shy around the girls. I didn't get much conversation out of him more than "Hi." It was amusing, but seriously, I'd have liked to hear him talk about how he used programmer code to make Fred move.

But mostly, I felt *huge* relief that Fred had a programmer and we were still in the game!

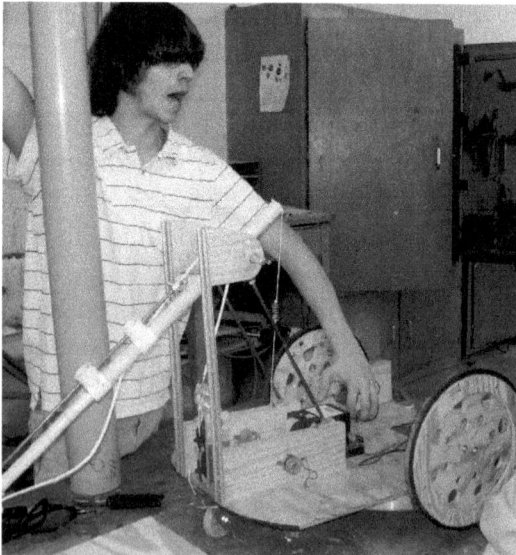
Broghan during robot construction

FOURTEEN

+

PAUL

Waiting for Gibby find someone to program Fred's brain about made me crazy. The Hub competition date grew closer each day. We needed that programmer! It wasn't Gibby's fault, so I didn't blame him. I was just frustrated that there wasn't anything Phillip or I or anyone in the Conway community could do to help.

After striking out with professors and grad students, Mr. Gibson again called the MSU Computer Sciences Department Chair. He said there might be an undergraduate student who could help. *An undergraduate? We go from having a professor help to an undergraduate?* Gibby must have been skeptical but he didn't say so to any of us at CHS. He probably told the professor something like, "I'll take any help I can get right now. My kids can't go on until that robot's brain is operating."

When Gibby got Henry Stratmann's email offer of help, he did fist pumps in the air.

The rest of the club learned about Henry when we gathered after school. Mr. Gibson turned club time over to Mr. Coyle and took Fred to MSU to meet Henry.

FRED 2009-2010

FIFTEEN

HENRY

When the CSI Department Chair explained Conway High School's robotics project and said they needed someone to program the robot's brain, I smiled. None of the upperclassmen in our club was interested so I raised my hand. The chance to program a robot was a major challenge and I was, to say the least, intrigued. I'd been a gamer all my life and a chance to program a robot and watch it work in real time, not virtual time, was super cool. *Bring it on.* The chair thanked me, handed me Mr. Gibson's contact information, then went on with the meeting.

Right after the meeting, I sent Mr. Gibson an email message offering my help. He replied immediately. We set up an appointment for him to bring the brain to MSU and explain what he needed. He was anxious to get going so he brought the motorized robot body with the empty brain the next day, soon as he could away after his classes. We talked about the basic moves Fred needed to do on command so the robot drivers could start practicing.

"The program code is on the BEST web site," Mr. Gibson said. He gave me his password so I could log on and download the code. "I expect it's a massive amount of data. I presume you have a high speed connection that can handle that?" he asked.

"Sure. Probably no more code than I've been using with homework. I'll work on this at home tonight and I'll update you tomorrow," I said.

He also gave me some DVDs with some of the data.

"Email me as soon as you have something. I'm stuck. Can't do this level of programming myself. I have a little training but—" He stopped. Shrugged his shoulders.

"I think I can help you," I said. I could feel his pain. Mr. Gibson smiled and pumped my hand, then thanked me again.

I wonder if he slept much that night. I doubt it. He's pretty intense and he loved robotics and those Conway kids. No worry for me. Programming Fred's brain would be mere brain-fingers-eye coordination.

Right after supper I got onto the site and started the download. It was a huge amount of code, but nothing my equipment couldn't handle. I connected the robot's brain to my computer and each time I downloaded a new component, it took about 20 seconds to test. Seconds added up to minutes, then hours, but I was fueled with PowerAde and determination. Shortly after midnight I'd accomplished what Mr. Gibson asked and I emailed him. "Fantastic," he replied.

He collected the brain that afternoon.

Mr. Gibson and the main robot builders, Paul Coryell and Phillip Foust, came to MSU a few more times to add moves to their robot. They were good guys—and smart. I really enjoyed working with them.

I went out to Conway a couple of times after school and one Saturday, too. For one thing, I wanted to see their school. They had a sixties era brick building with the shop tacked on to the back and a whole bunch of portable buildings attached to the school. No frills. The shop equipment was really basic and some of it didn't fully function. Some chair seat covers were ripped, and the metal parts were all beat up.

They had ancient computers in their lab so no wonder Mr. Gibson couldn't download the programming code. It was also obvious the school didn't have any money to hire a programmer. Because of all that, I was surprised and impressed with the level of their robot's sophistication.

I later found out that some of the Conway kids thought that I looked like Harry Potter. That was funny.

I was super glad to help them and most of the time, it was a lot of fun. Eventually, ironically, my picture with the Conway robotics team and Fred ended up on a huge poster in the MSU Engineering building. Sweet! But before that, we had to MacGuyver some things, and we sorta broke into Cheek Hall one time. We needed to use some equipment early one morning and found a door open that probably should have been locked, so we went in. But most of all, we had a whole lotta fun making Fred into a working live action droid.

Phillip, Paul & Henry

SIXTEEN

+

PAUL

While Henry programmed the basic robot moves, Phillip and I kept working on the arm. It needed to retract because BEST rules required that all robots and attachments fit into a box, a two-foot cube.

I had trouble figuring out how to make the arm reach out and then retract. We had another group meeting. Some guys think that if they get louder their idea will sound better. Didn't happen. I came up with the idea of using a rubber band. It worked.

One problem solved.

Making the rest of the arm about made me crazy. I couldn't figure how to make the end pieces open and retract enough to grab the hanging beach ball-type globes, and it was really frustrating. That arm gave Phillip and me so much grief that I actually thought about giving up. Quitting isn't something I do easily, but that problem about did me in. That's the only time in all my years in robotics that I almost quit.

We eventually crafted an aperture for the arm that had circular pieces that could open enough to grab hold of the plastic globes hanging above the field and close to get the smaller balls and cans.

The controller provided by BEST was a four-function device consisting of two joysticks. One joystick controlled up and down motion, and the other moved the robot left, right, or forward.

The controller operated by a battery. We purchased a second

battery because the rechargeable batteries supplied from BEST, like the motors, had been used a lot and wouldn't hold a charge for more than a couple hours, then required twelve hours to charge.

After worry about the robotic arm construction, it was worry about programming the brain until Henry saved us.

Sometimes Henry and Phillip argued when the droid didn't move the way we wanted. When that happened, it was usually a hardware problem—our error, not Henry's programming. Phillip would bring out the screwdriver and pliers, fiddle with wires, and try again.

I usually drove Fred. That is, I manipulated the joystick controller. The first time Fred responded, we all went crazy happy.

"He's moving!"

"Go Fred!"

"Yessss!"

Fred rolled out the open shop door, then I made him turn and zoom back inside. *Wow!*

After Fred had a brain, it was time to set up driver tryouts. I told Ceira that even the girls would get a turn. She hit me on the arm with her fist. Hard!

"Ouch!"

"You had that coming," she said.

Yeah, I did.

FRED goes out the shop door.

Hopper

SEVENTEEN

+

CEIRA

A ton of people showed up for robot driver tryouts. Some came to watch, but most were driver wannabes.

Tryouts had the energy of a bull ride, the body moves of basketball, and the rhythm of marching band practice. At first, tryouts didn't include the math and chemistry calculations needed in competition, but that was no prob. Most of us in Robotics Club were good math and science students.

Sara Carnes and I were determined to break the male monopoly on driving.

First rounds of driver tryouts focused on who could maneuver the robot to gather the most cans, balls, and globes they could and dump or shove the objects into the corner of the game field in three minutes. That translated basically as who had the best coordination *and* most aggressive gaming skills.

The robot's controller was like an oversized X-Box controller and I'd used one of those a gazillion times. In fact, most of us had wiggled joysticks soon as we were out of the cradle. Only thing, the guys had the most aggressive gaming attitudes. They slammed that robot around the field and jammed more stuff into the corner than either Sara or me. We focused on accuracy *and* speed. Mistake.

At the end of tryouts, a gender mix wasn't part of the equation; testosterone ruled. Instead of being a whiny girl, Sara and I stepped aside and carried on with corporation stuff.

Given more time to practice I could have done it, but the first competition was just a couple weeks away. Gibby was anxious to get the driver-spotter team honed to six, so no one got more than one chance in the first round of tryouts.

BEST rules required a team of student driver-spotters, as many as six, for competition. No one could be the sole driver of a robot; they had to rotate the driver position at the beginning of each round. However, during actual competition, as robots moved into semi-final and final rounds, the fastest drivers and most efficient spotters would rotate into the position every other round. That's when the crowd really got into the action yelling "more water" or "more CO2" or whatever item was needed to up the score that round.

The most skilled drivers were the rock stars of BEST robotics competitions. Honestly! Some even dressed like rock superstars with body paint, wild hair styles and colors, and flamboyant clothing in school colors. Not Conway. We weren't flamboyant. Leave the wild hair to other schools. We were more into being lively nerds, not mad scientist types.

It was an interesting mix of characters that made Conway's final team—guys who didn't ordinarily hang out together—juniors Paul, Phil, Broghan and the three seniors, Shane, Lloyd, and Grant.

Broghan was the best spotter, but he did okay as a driver, too. He just had to keep his grades up. Babysitting interfered with his homework and he'd gotten perilously close to academic probation. I was really glad he made the team. We were talking a lot by then and the more we talked, the more I liked him for his personality, his ethics, and his sense of humor.

Back to working on the table display for me.

One afternoon, I had a huge hunk of cloth I was preparing and needed to use a sewing machine. Paul brought in an old sewing machine from home that he'd bought at a garage sale. He set it up on a table in the middle of the classroom next to the shop. The machine worked pretty well except for one part that drove the wheel. I had to use one hand to turn the wheel and the other to guide the cloth. It was tedious, to say the least.

I was slogging away one Saturday and turned away to answer someone's question, when all of a sudden the wheel started moving fast. I heard the unmistakable *zzzzziiinggg* of a hand held drill. Paul held the drill to the sewing machine part that make the wheel work and I finally got that piece of cloth hemmed. Teamwork!

The week of the Hub competition, we had our new and very sharp t-shirts, along with pompons, noisemakers, the web site, and table display stuff ready to go. Those of us who were the faux corporation officers had our speeches memorized.

Then someone dug out the Bubba Bear mascot costume. It was nasty! That thing was hot on the bod so it was infused with old sweat. Yuck. Someone finally took it for dry cleaning and that helped, so several people volunteered to take turns wearing the thing.

Our spirit team would wear our snazzy t-shirts with jeans or khaki pants and comfy shoes, except for the business clothes we had to wear for the corporation team's oral presentation. No black-white-red streaked hair or wigs for us.

We were pumped for the first competition of the High Octane year. We felt pretty confident that we'd go on to Regionals but you never knew which schools had really dug in and worked on their robot and which school had practiced most. So there was still some tension as we packed for the Northark Hub.

Classroom Practice.

Ashley and Bubba Bear.

EIGHTEEN

+

CEIRA

"The bus leaves in five minutes! Grab your stuff and get going or find your own way to Harrison!" Mr. Gibson's bellow made thirty club members jump to action.

We had three hours of bus ride ahead of us to get to Northark College. The realization that six long weeks of work would soon be judged in live competition suddenly hit with Gibby's yell.

Teresa Rumfelt, Grant's mom, brought their truck and a trailer to haul stuff that was too large for the bus. Twelve three- foot by eight-foot side panels, three folding tables, the display top, shelves, tool boxes and other stuff went into the trailer.

Mr. Coyle drove down alone, as usual. I didn't blame him for not riding the bus since he was semi-retired and didn't have to be with the club.

Mr. Gibson rode the bus with us. As our sponsor, he officially had to, but I think he liked the craziness of being with us. Or maybe he just acted that way to get us pumped up for the competition.

Fred in his travel crate, wouldn't go through the front doors of the bus so he was loaded through the back emergency door. The crate sat in the aisle, barely fitting between the seats. If we had to make an emergency exit, we'd have to climb over the crate to get out. Probably an illegal setup but what could we do?

We crammed our back packs under the seats, dug out cell

phones, games, and books and were on the road to the first robotics competition of the year.

Not far into the trip someone yelled, "Who's got food?"

My response was to throw a granola bar toward the voice. I'd packed snacks because some of the club members didn't have money for the concession stand.

The ride passed.

The Northark College field house doors opened to blasts of AC/DC. The robot boys headed for the pit area, the spirit team installed themselves in the stands and Shane went looking for cute girls. I got my presentation team together for one last practice— Elizabeth Deckard, Chris Cook, Shyanne Witt, and Braden Jones.

HUB

NINETEEN

+

PAUL

As we were the setting up the pit stop table, I heard someone say, "Oh shit! Where's the backup battery?"

I looked. "Not here."

"So it got left at school!" Phillip said. "Who can we call to bring it to us?"

"We've got to get this covered before Gibby finds out. He'll kill us for sure," I said.

We were in a major panic. We called Colton, who was driving his car down. He was already past Springfield and more than halfway to Northark—too late for him to turn around.

"Ah man, we've gotta tell Gibby," I said. I'd rather have eaten nails.

Mr. Gibson went ballistic. I wasn't happy, either. We all felt bad.

"I'll call Lisa," he said. "Hope she hasn't left the house yet."

While he called, lots of hissed finger-pointing went on behind his back. No one would take the blame.

Mrs. Gibson was driving down in their car. She agreed to go to Conway to get the battery and bring it to Harrison.

Although the problem was solved, Gibby continued his rant.

"I thought you had checklists of things you needed to bring!" he yelled.

"We do," I said.

"So who didn't check off the battery?" he snapped, hands jammed on his hips. Chin out.

"Not sure about that," I said. No one argued with me that time, not even Shane. In fact, most of the team had backed off and stood with arms crossed, trying to look innocent. Phillip stood with me, but didn't say a word.

"You'd better get your act more organized than that if you don't want to make fools of yourselves," he yelled, and then stomped off.

As if we didn't know.

TWENTY

+

CEIRA

Paul and Phil finished setting up the pit area then came over to help construct the display booth. The pre-drilled holes in two-by-fours made it quick to assemble. I took my turn with a drill and hammer, then put the chemicals on the shelves.

In the midst of setup, I took some still shots of the process. Last, I made sure the PowerPoint worked. It would repeat whenever someone touched the screen so that people could visit the display any time during breaks in the competition. The corporation notebook would remain on a table in the booth, and a larger table held my mock chemistry lab. We pinned posters to the walls to complete the display.

Judges visited each booth, armed with clipboards and checklists. We'd get the results of their decision later in the day, when they announced the BEST award winner. First, we had to do our corporation team live presentation, be spirited *all* during the robot competition, and show off our t-shirts.

At 8:00 a.m. the announcer interrupted the head-blasting rock music and introduced schools. I looked around. With thirty people, we had the largest group there. What a novelty for Conway!

Our neighbor school in Lebanon, Missouri was much larger, but robotics there was part of the gifted program so they didn't bring a large group to competition. They had more parents there than we did, though. It wasn't that our parents didn't support us, but they had so much to do on weekends that not many of them could afford to take a Saturday off for robotics.

There were a surprising number of college people and local townspeople in the Northark field house and I could see that local media covered the event. Word must have gotten out about how much fun the robotics competitions were.

Crowd excitement was high as any basketball tournament. 'Bout time!

The smell of popcorn and coffee floated down from the concession stand as I took my team upstairs to the mezzanine, then outside and into the classroom building next door. Oral presentations took place there, while robot competition started on the field house floor.

We found a restroom and changed into our business clothes, then waited our turn. Even though I'd done this gig before, it triggered sweaty palms and a jittery stomach to be waiting our turn for judging.

Ceira at the computer with onlookers.

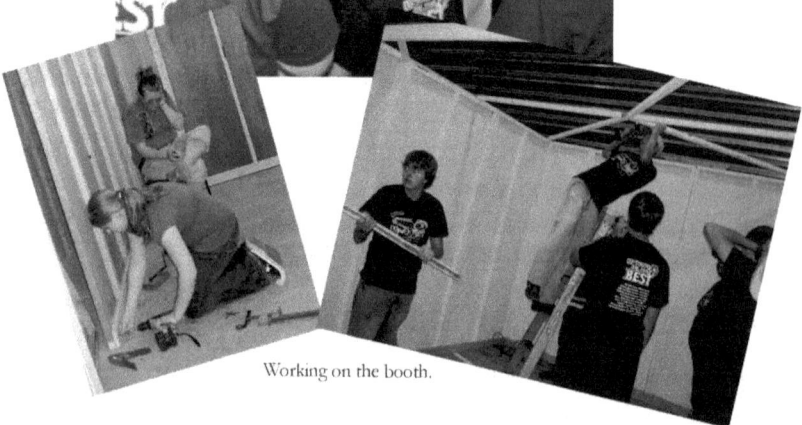

Working on the booth.

+

PAUL

On the floor of the field house robots and teachers lined up for the judging table. Each robot was weighed, measured, and examined for illegal parts. Next, the robot went into the two-foot cube for measurement. Fred passed, no surprise.

Gibby stood close by, rule book in hand, chewing gum at full speed. He was known for his knowledge of the rules and regulations. In fact, Mr. Gibson had such a good rep that sometimes the judges consulted him in the middle of the competition and he'd flip to the appropriate page in his three-ring binder for reference.

The official game field was a twenty-three foot square. The square was divided into four different colored quadrants. In the center of the square sat the commodities dispenser, officially called the Energy Generating Station. Four different schools' robot driver-spotter teams occupied the four quadrants during each round of competition.

When the starting buzzer blared, the dispenser would dump cans and balls and they'd roll into all quadrants. Instantly, drivers sent robots after the items. Fortunately, we'd had practice rounds and found out where the minor bumps in the carpet were because Fred's scoop was close to the floor and he'd stall on the bumps. Each driver had practiced navigating Fred around the bumps.

Gibby reminded us of the value of each item, then he went to the stands, as required. Teachers could not help in the pit area. The pit was student territory—and judges territory.

On the field, each six-ounce sized tomato paste can represented one unit of energy. Each racquet ball represented one unit of H_2O - water. Each yellow tennis ball represented one unit of catalyst. Each beach ball type globe represented one unit of CO_2 - carbon dioxide.

During the elimination rounds, the drivers collected only water, catalysts, energy, and carbon dioxide. The spotter only had to direct proximity to the driver when to dump the load in our corner refinery.

We were in the third group, so watched our competition. Some of the other schools seemed like they'd barely practiced, and could hardly maneuver their robots. Maybe they didn't have a mock practice field, like Mr. Gibson built for Conway.

Our first round was a bust. When the buzzer signaled "*GO*" Fred didn't move. I could have died right there. Wanted to just melt into the floor. We'd forgotten to turn on the robot! A simple switch. *GAAA*! I didn't dare look at the stands because Gibby's glare would have been a killer.

Fortunately, each team is allowed to throw out one of the rounds. That first round was ours.

From then on, thankfully, everything worked like a dream. Every driver did well. Broghan was the main spotter and he was excellent. We cleaned the floor of most all commodities, the cans, balls, and globes, leaving little for the slower competitors.

Elimination rounds ended with us as top dog by a huge margin. That felt good. Really good!

Next came semifinal rounds where the chemical equations created by the collection of cans, balls, and globes had to work to create isooctane. The driver and the spotter would calculate in their heads while they guided Fred to collect objects. Each three-minute round got our adrenaline going and our brains focused on chemicals.

After each round, we had backup from Gibby and his spreadsheets. One of our team members would go check with him to see what the spreadsheet showed we needed for the next round.

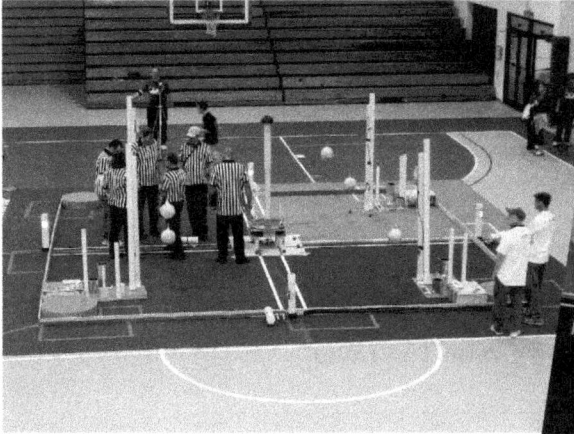

Getting ready.

FRED retrieving globe.

TWENTY-TWO

+

CEIRA

At 8:35 A.M. our corporation team did our thing in front of the engineer-judges. I mentally fought the discomfort of pantyhose and heels and tried to act calm and professional. We took turns as we explained the process of robot design to completion, using our PowerPoint slides to highlight main points. No one flubbed lines although voices quavered a bit.

Even though our final practice had gone without a glitch, it was nerve-wracking to do our talk in front of the judges. We finished, said our thank yous, and kept our poise as we left the room. Actually, I sort of teetered out of the room on my high heels. Right outside that room, I chucked those things. All of us rushed to the rest rooms to change into our team t-shirts, jeans, and tennis shoes. My toes went *ahh*.

We were anxious to see how the robot boys were doing.

Inside the field house, noisemakers of spirit teams from eight schools competed with deafening rock music while the floor action rolled on. We jogged down the stairs to join the CHS group and yell for Team Fred.

We were in first place!

"Here we go Conway, herewego!" *STOMP STOMP* We loved being the biggest school group at the Hub. We chanted, clapped and cheered from the stands as Fred whomped the competition.

We could be heard even over the blaring rock music. Some of

the adults used ear plugs. Wimps.

Sara Carnes had painted Fred last thing the night before competition. Her design was fabulous. The stadium lights highlighted Fred's shiny red and white paint. He out-shined the other robots, hands down.

"Go Paul!" My robotics brother was driving. Broghan was spotting. He was a great spotter–no guessing about his hand signals. He could have been an airport flag man guiding planes to the terminal. The robot driver had blind spots where he couldn't see when it was right to dump the hopper so the load didn't go rolling around the field again. Broghan was right on.

Each of the drivers did well. The strange combination of personalities of the six guys meshed on the playing field and they became one team.

Amazing.

The music switched to the YMCA song and the crowd danced. Cheesy music was part of the BEST robotics tradition.

Judges in black and white shirts patrolled the game field perimeter, watching the action of four robot teams competing for parts. A robot could go into another school's quadrant and nab commodities, but if that robot tapped your robot, you had to freeze in place for 30 seconds. That usually meant a losing round for the frozen robot.

Paul and Broghan completed a good round and our score went up.

We were way ahead. Our guys were making High Octane like crazy. Those three-minute rounds went fast for those of us in the stands. But the guys driving and spotting said each round seemed like an hour.

Shane was the best driver on our team for stealing others' stuff. I mean, he was brazen. He loved to taunt the other team by driving Fred toward their field, back, toward again so the other driver got distracted. Having his two best friends on the team, Lloyd and Grant, energized him. He was good before, but he'd become a super power in live competition.

The scoreboard buzzed the start of another round. *Oohh.* One middle school's robot was D.O.A. *Sorry guys!* They carried their dead robot to their pit table and dismantled the parts that had to be returned—motors, brain, servos, controller, and the other electronic stuff.

The CHS spirit team added "Dread the Fred!" to our chant repertoire.

Broghan headed to the stands and I caught his eye. He smiled. He was so cute when he smiled!

Ceira's presentation - HUB

TWENTY-THREE

✦

PAUL

Teams that didn't qualify for semifinal rounds dismantled their robots and left the returnable parts with the judges. Some seemed sad and others seemed relieved. We were polite and told them "Good job," "See you next year" and stuff like that.

Scoring for the High Octane game was way more complex than in any prior year's competition. Each team carried over inventory from round to round. For example, if we had one benzene, three water, one energy and one catalyst accumulated, the driver could mess up if he gathered a water unit without also gathering chemicals to produce ethylene. If he did that, the reaction for naptha would be calculated in the scoring program and we'd have to gather more benzene and try again.

The spotter could discard an unwanted water unit if it kept the team on track. That's why we had to check with Gibby or whoever was tallying our scores in the stands. I should mention that each game piece– alls, cans, globes–had assigned receptacles in our corner area. To make it more complicated for semi-final rounds, they added a benzene tanker that the robot had to maneuver into the refinery. Proper use of the tanker added points.

When the driver maneuvered the robot into position, the spotter could use his hands to put items the robot dumped into various receptacles. The spotter was the only one who could touch items during each round. He had to be fast, and accurate. His body, however, could not be on the game field.

There was so much for drivers and spotters to do in a three-minute round of High Octane!

Drivers could maneuver their robots to block each other, and could travel into any of the four quadrants of the game field to gather commodities. However, you couldn't use destructive behavior. That is, you could not deliberately slam your robot into a rival robot.

You couldn't be shy with your robot. But, if you drove your robot into someone else's territory, you had to keep an eye on where their robot was, as well as the other two robots. You didn't want someone to scoop up loose chemicals from your section.

We were way ahead in the semi-final scores when the announcer blared out, "Lunch break. We will resume action in one hour."

Mr. Gibson was almost embarrassed at how far ahead we were. I wasn't. I was relieved and proud. I knew then that Fred was a good robot and we had a great team of drivers and spotters.

Mr. Coyle smiled his shy smile. He stood in his usual pose with his hands behind his back and his belly out. I heard him tell a parent, "I'm just happy to see kids excited and enjoying the application of math and science in a real world event. I don't care about scores. Every kid here is a winner." He was sincere about that.

People stormed the food stand.

I was starved. So were the other drivers. A morning of competition really burned up our energy. Someone from our group had gone for pizza and man, that tasted *so* good!

Lloyd & Broghan with FRED

The judge & FRED

TWENTY-FOUR

+

CEIRA

The Bubba Bear costume rotated around to whoever could stand to wear the thing. The girls gagged at the smell of it but most of the guys didn't seem to mind—not that they'd admit, anyway. I pulled a can of air freshener from our supplies and sprayed BUBBA inside and out. We'd let it air dry through lunch.

After lunch, we kept building our score. Fred and all other robots went to the pit area after each round.

At one point, Shane attempted an adjustment on Fred and Paul yelled at him. Gibby laid down the law. No one but Paul and Phil could tinker with Fred. Shane didn't argue much that time–he had girls to flirt with from all the other schools.

Our guys took turns as the robot driver and field spotter, although Broghan did more rounds as spotter than anyone else. He also helped Mr. Gibson some with the laptop score.

As usual, the robot competition took longer than planned and it was late afternoon before the four top-scoring robots were ready for the final rounds. As the day went on, we'd steadily gained scores. The spirit team got louder. We ended semifinals in first place. Feeling good!

The scoreboard went blank and the announcer took the mic. The rock music stopped. For a few seconds, the stadium was silent but our ears kept ringing.

"Ladeees and gentlemen! We will begin the final four rounds of competition in ten minutes. During the final rounds, no scores will be posted. We will not announce any scores until the first, second, third, and fourth place robot teams are ready to receive their trophies. We will also announce the winner of the BEST Award at that time."

The announcer spoke as if all was wide open but everyone knew that Conway was going to take first in the Hub robot competition, barring some catastrophe. The only mystery was who would be second, third, and fourth.

Broghan and the rest of the team headed for the stands.

I was anxious to score well in our corporation parts, too and hoped for the BEST award so that Conway, for once, would be the overall champions. We'd been in BEST robotics events for a few years now, and sometimes we placed high overall, sometimes not. Sometimes the corporation component did way better than the robot and in other years, the reverse happened.

This year was going to be a first place robot trophy, no doubt about that. I was so happy for Paul. He and Phil were such great guys and Fred was awesome!

At the end of the four final rounds, the judges went into a huddle for final score tallies.

Awards and photo op time were always a high point and a great way to end the competition.

Broghan sat next to me! I almost fell out of my seat.

Add to that my happiness that our notebook, table display, and oral presentation all receive good evaluations. We won the BEST award!

Somewhere during the trophy presentation to the third or second place robots, Broghan put his arm around me. My heart almost leaped out of my body!

Finally, it was photo time for the first place robot team and the entire CHS spirit team thundered onto the floor.

First, they posed the driver-spotter team and Fred. I took still shots and some video of the six guys standing behind a table where Fred and the trophy proudly rested.

Awesome.

In the photo of the robot boys with Fred and the big winner trophy on a table in front of them, you can tell a lot by their posture. They stood shoulder to shoulder, but in two groups. Grant, the tallest, is on the far left, then Lloyd, Broghan and Shane. Shane stood with feet further apart than anyone with hands clasped in front. He and Grant weren't smiling. I think they tried to look macho. Lloyd and Broghan have small smiles. There's an obvious space between Shane and Paul. A huge grin lights up Paul's face. Phil is on the right, next to Paul. He has his hands behind his back, his cheeks are bright pink, and he is smiling big, showing his dimples.

Last, the entire club lined up for photos, and I joined the group. Paul's mom took digital photos and video that I'd use for updating our stuff for regionals.

The announcer closed the evening with information that regionals would be at the University of Arkansas at Ft. Smith on December 12. The top winning robots from eight states would be competing.

Eight states! More than last year. Are our guys up to it? Am I?

On the bus ride back to Conway, we gossiped, played games, listened to music and napped. I tried not to think about regionals It was just so damned good to celebrate this Hub victory!

Lloyd, Grant, Broghan, Shane, Paul & Phillip

Ceira with HUB BEST trophy

TWENTY-FIVE

+

PAUL

What a relief! Our big win meant Fred was an awesome robot and that our drivers and spotters could work well together in competition even if we didn't hang together otherwise.

Ceira, Sara, Shyanne and all the rest did well with the corporation stuff to win the BEST trophy. That was super. But regionals would be much more intense and we'd compete with some super sophisticated robots and professional-looking corporation presentations. I'd been there. Some of the teams at regionals had aerospace industry helpers. We'd be the smallest school there.

The Hub win gave me confidence, but I didn't get over-confident. The action in competition showed me that Fred needed some work to make him even better.

On the bus ride home, Phillip, Gibby and I talked about some things that I wanted to change in Fred before regionals. The wheels needed to be larger for smoother and faster operation, and we could make some adjustments to the body so the robot would weigh just a bit less, which would also help with speed on the floor. We'd be back in the shop next week.

I'd have to do some research on wheel size ratio to torque power. After that, we'd just need to practice more so that each driver got better and we'd be ready for the bigger and more intense competition at regionals.

Eight weeks until regionals. Lots to do.

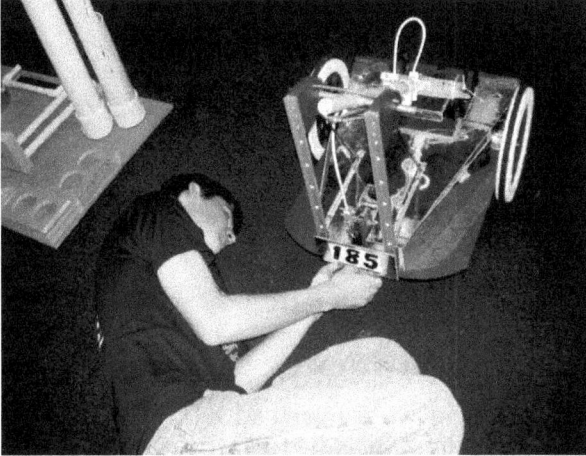

Paul adjusting FRED

TWENTY-SIX

+

CEIRA

We were pumped. We savored our victory like no other because it was the biggest win we'd ever had. Never before had we been the school with the loudest spirit team *and* the snazziest, jazziest robot. Never before had our drivers and spotters kicked butt like they did at the High Octane Hub. Never before had our corporation team and table display been so lauded, and I was the CEO. Phillip paid me a fantastic compliment when he said that I'd done more than the teacher who helped the corporation. Awesome.

With some reluctance, I thought about regionals. I knew that regionals would be a much, *much* tougher competition, where the private schools with megabucks for resources would outnumber public schools.

Between hub and Regionals we got into the groove of school work, sports teams, and band. Only hard core robotics people gathered after school each day. We added pictures and descriptions of the hub competition to the notebook and the web site. I pretty much kept the table display the same with the mini Chemistry lab. Gibby criticized my poster for the table display booth but didn't offer any concrete ideas or materials for how to make something that he'd actually compliment.

"Some days I want to hang that man up with a chain on his heels and skin his hide with a rusty butter knife." I vented to Paul and Phil.

"Hey, at least he doesn't call you Doofus," Paul said.

"Or Dingus," Phil added.

I grinned. "Well, yeah, you've got me beat there."

While Paul and Phillip rebuilt some parts of Fred that had been awkward in competition, I went to the shop quite a bit to watch and talk with my best friends. It was where anyone seriously into robotics went to hang out. Conway didn't have any fast food or places where you could hang with your friends. The two quick-stop-gas-n-guzzle places had a couple of tables, but old men with coffee dominated those. It was too expensive to drive to Lebanon or Marshfield for fast food so we did robotics, and sometimes we watched videos at someone's house.

Broghan was now very talkative to me, and he helped some with Fred.

Grant mostly went to basketball practice while Paul and Phil worked on Fred. Lloyd and Shane hung out in the shop some, but they weren't engineers so didn't work on the rebuild very much. After Grant finished practice, they'd all go somewhere to play games and eat mass quantities of whatever snacks they could bum. One time they stayed out all night doing only God knows what and they told their parents they were in robotics. Even though Grant wasn't my favorite person, I wasn't about to rat him out!

One night after robotics, after I gave Broghan and his sister Ashley a ride home, she went inside and he stayed beside the car with the door open. I looked at him. He looked at me, wasn't smiling and I wondered what was going on. My tummy clenched. All of a sudden he asked if I would go out with him. *Finally!* I said, "Yes!" and smiled.

He smiled, too. Broghan and I were officially a couple.

As the semester developed, Phil and I had some conflicts between band and robotics. Band practice focused on getting ready for Christmas parades plus the school Christmas concert. Christmas parades in Southwest Missouri were hugely important. We knew about Kwanzaa and Hanukkah, of course, but didn't have any families in Conway who celebrated those holidays. The diversity in our community came with churches. You went to church or you didn't. A mixed marriage would be one between a

Baptist and a Methodist. Or a Catholic and a Protestant.

Our CHS band director thought that Christmas parades were more important than robotics so I had to stand my ground. I explained to him that robotics was an academic event, and super important for those of us who valued academics and were headed for college.

I was the only trombone player so when I wasn't there, the low brass section of band was weak. Phil played percussion, and without a strong percussion section, the band didn't march well. Phil and I understood the band director's situation, but although we loved music, we valued robotics more.

We didn't miss any parades until Regionals, and then all hell broke loose.

Band - Ceira on trombone & Phil on percussion

TWENTY-SEVEN

✝

PAUL

To get ready for Regionals, Gibby pushed us hard to do better in driving the robot more confidently, and to gather chemicals faster. He wanted each of us to maneuver Fred without a second thought.

He gave each of us a print copy of the rules and laid down the law. "Learn the rules! He ordered. He knew we'd sluffed by for the Hub but that wouldn't happen at the next level. He wanted each team member to know when to challenge a judge because sometimes they made mistakes.

Phillip scanned through the rules at least once. I sidestepped detailed rule reading but listened when Gibby drilled us, especially during practice. It just worked better for me to apply the rules during action.

BEST guidelines for calculating High Octane inventory:

1. Commodities deposited in appropriate receptacles will be tallied with commodities already in inventory from previous rounds.

2. If sufficient quantities exist for a stoichiometric balance then the reaction producing the highest possible product will be exercised first.

3. If sufficient quantities remain to satisfy other reactions, the remaining will be exercised with the most valuable first.

4. Once a commodity is reacted to produce a product, it is removed from the team's inventory.

5.   If the remaining tally of any one commodity type is four or more, then four units of that commodity will be traded for one unit of the next more valuable commodity.

6.   Steps three through 5 are iterated until there is no further change in the team's inventory until the next round.

Gibby translated. The first section of rules referred to the chemical equations made each time we grabbed a tennis ball, racquet ball, tomato paste can, or globe.

More rules:

Each game piece has assigned receptacles:

1.   As many as five H_2O units in the specified linear trough on the floor of the processing cell and one H_2O unit may be placed in each of the vertical pipes atop the storage area.

2.   As many as 5 catalyst units may be placed in the elevated linear trough in the processing cell; 3 units may be placed in the linear trough above the storage area.

3.   As many as 5 energy units may be placed in the "L" shaped receptacle located in the corner of the processing cell; one may be placed in each of the three receptacles atop the storage area.

4.   One CO_2 unit may be placed in the lower circular receptacle and one may be placed in the elevated circular unit.

5.   As many as three benzene tankers may be placed in the tower level of the storage bays.

The driver, the person BEST called the "field specialist" had to steer the robot to the dump zone for the correct receptacle.

In general, it is the operator's duty to relocate items from the offload area in the center of the 4-quadrant game field into the home base storage area. The operator must also give non-vocal guidance to the field specialist as appropriate."

Gibby's translation: The spotter was the "operator." The spotter learned his role during practice. He had to be accurate, and fast as possible. Broghan was a good spotter. Shane was the best. The rest of us were okay spotters.

Operator guidelines:

1.   The operator's entire waistline must remain within the boundaries of the area, a 2' x 5' area outside the field perimeter, or atop the circular operations platform. The operator may extend arms or legs outside the areas as long as no other rules are violate in doing so.

2.   The operator may relocate any game piece or robot subassembly that lies entirely within accessible areas.

3.   The operator may not use any artificial tools for communication or game piece manipulation.

4.   The operator may not touch CO_2 units.

5.   The operator may not touch or manipulate Benzene tankers

6.   The operator may not touch the lower storage area or any item that is at least partly within that area.

7.   The operator may not touch the field floor or boundary except within the operator platform and offload area.

8.   The operator may not contact any game piece or subassembly that is also in contact with any robot.

9.   If an item is accidentally moved outside of the operator accessible areas the robot must move it back into the area before the operator can handle it.

10.   An operator may not use any item as a projectile.

11.   An operator may not cross the boundary of a processing cell. An operator may roll items into any processing cell area.

12.   An operator may return game pieces or subassemblies back to the playing field by dropping them or rolling them.

13.   If an operator's actions damage another robot, the operator's team may be disqualified whether the damage was intentional or not.

Additional rules:

Deploying a blocking device on the game floor is not considered destructive behavior if such a device is passive in nature, even if another robot damages itself by interacting with it.

Nudging another robot is not necessarily destructive behavior.

Inserting your robot arm into the electronics area of another robot where unsecured wires might get unplugged is destructive behavior.

Each match shall be three minutes long and is played with a maximum of four teams. If necessary, matches may also be played with fewer than four teams. The scoring software will assign teams to a match and will determine the teams' starting locations relative to the game field.

Gibby didn't know how judges would decide the difference between a robot "nudging" another and when it was using too much force so he told us to not let Fred touch another robot at all, if we could avoid it.

Then Gibby referred to the part in the "Additional Rules" section that allowed for a blocking strategy and challenged us to devise something.

We dug around in the parts kit and came up with a roll of duct tape. Someone ripped off a hunk and rolled it into a ball and suggested that we roll the balls onto the game field to block opponents' robots. Someone else said the duct tape wouldn't last long that way, but if we used some cardboard for the center and wrapped tape around it, we could get balls large enough to make a robot stall, but not damage the robot.

A wild argument erupted when Broghan said that he didn't like it and didn't think it was a fair thing to do.

Mr. Gibson's rationale was that the rules were stated such that the engineers of BEST challenged schools to be creative. Making duct tape covered balls of cardboard was creative. And we wouldn't damage a robot with them, we'd just slow them down.

He reminded us that we were headed to a much more difficult competition at regionals because of the bigger schools having the resources to build very sophisticated robots. Wichita Homeschool had huge advantages. They built the actual field used in regional competition, and they used it for their practice ahead of regionals. They also supplied the judges for regionals.

"I'm just trying to level the playing field," he said. "They have more money to build fancier robots but we can be more creative."

Broghan stood firm on his principles. He argued that the blocking strategy wasn't good sportsmanship and he himself would not use them.

Mr. Gibson lost it. "If you're not going to support the team, then you're free to leave," he said.

The way he said it was clear to everyone. He wanted Broghan to clear out.

Rebuilding 2009-2010

TWENTY-EIGHT

+

CEIRA

I was devastated when Broghan told me Gibby wanted him off the team. I agreed with Broghan's reasoning. It wasn't good sportsmanship to use the duct tape balls to block other robots. Broghan had to live by his principles, and I loved him for it. I thought seriously about leaving the team, too. But Broghan didn't want me to leave just for him. I was torn. Paul and Phil had worked so hard and they had a great robot and I had many other friends in robotics, not the least being my corporation people.

I decided to stay.

Broghan kept coming to robotics after school and he helped Paul and Phil in rebuilding Fred. He wasn't there just to goof off and steal snacks like some other people that will not be named. He also kept practicing as spotter, but he wouldn't touch the duct tape balls.

For regionals, our spirit team needed stuff to make more noise. Our vocal chants would be drowned out by the bigger spirit teams. If it went like the year before, we would be *the* smallest group there.

I bought several cans of Pringles, dumped them into a giant baggie for snacks, and we covered the cans with construction paper. Inside each can, we put a handful of dried beans. Noisemakers for little rural schools with no money for fancy stuff, but hey, they worked!

After Broghan *sort of* quit the team, Gibby really got crazy about watching us. I'm not sure what was in his head but we couldn't even hold hands or he'd yell, and you can't believe his reaction in the shop when one day Broghan kissed me on the cheek.

Gibby's mother-hen attitude really got nerve-wracking. But he saw that Broghan continued to help with robotics every day. He was there more than Grant, for sure. Eventually, he relented enough to let Broghan ride the bus to regionals and said he could take a turn at spotting.

Broghan liked spotting better than driving, anyway.

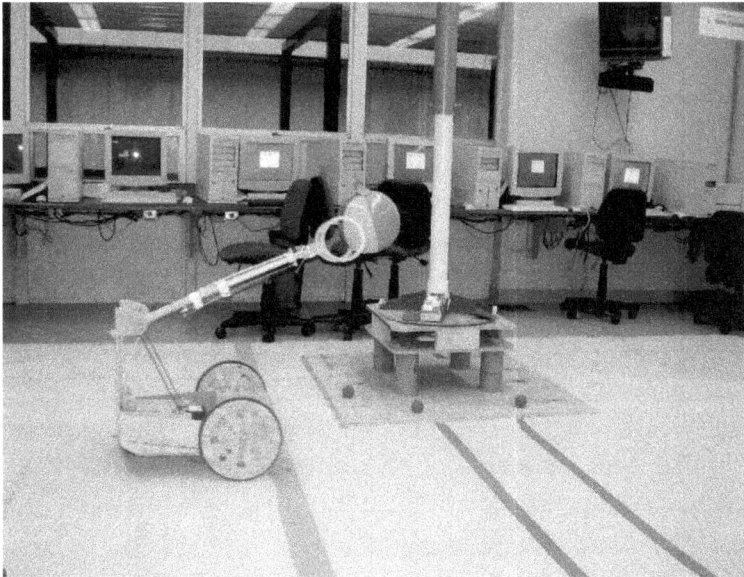

Practice with globe retrieval

TWENTY-NINE

PAUL

I was glad Gibby saw that Broghan was seriously helping Phillip and me so he eventually said it would be okay for him to come with us to regionals. That was fair. I could understand Broghan's point about sportsmanship, but the fact remained that blocking other robots was not a rules violation.

Besides, Gibby had kicked Shane off the team and let him back more than once. One time after a snow, Shane drove his car into the shop to let it thaw. Gibby told him to clean up the mess after the snow melted into piles on the shop floor. But Shane didn't. He left a big pile of slush that I helped Gibby clean up. Gibby was furious.

Another time, Shane was kicked out because he kept texting some girl during practice after Gibby told him to put the phone away.

I still needed to get more torque from the wheels. To determine the correct size of the wheels, I did some research on the internet. I had to figure the exact ratio of wheel size to motor torque and robot weight and eventually worked it out. Phil made the new wheels based on my specs, and they worked great.

Teamwork.

We, except Broghan, practiced with the duct tape balls to make sure our spotter could fling them into the other school's docking area. That way, the robot wouldn't be able to dock and unload chemicals, so their load wouldn't count. Shane was the best with accuracy in lobbing the duct tape balls exactly where they needed to go.

All of us laughed at the thought of blocking our arch rival, Wichita Homeshool and yes, that's the way they spelled their name. The private schools in robotics had huge advantages over the public schools. Not only did they have enormous budgets for robotics, they didn't have the restrictions on practice time that public schools did. Extra-curricular activities in public schools are strictly regulated to after-school hours. The private schools could work all day long for weeks and months on their robot activities.

Each year, we'd felt the differences grow until private schools dominated Regionals. The duct tape balls could level that playing field just a little—if others didn't have a similar strategy.

I was nervous, because this year's regionals would be bigger than ever, but I knew we had a good enough robot to finish in the top five if things went as they had the last couple of years.

THIRTY

✦

CEIRA

It was the kind of cold, clear December day that produced frothy clouds of breath as people shouted reminders while we lugged our stuff to the bus for Regionals.

I caught bits and pieces of voices but couldn't tell who said what.

"Did you make sure to get the extra battery?"

"Is the tool kit loaded?"

"*Extra battery?*"

"*YES AND THE TOOL KIT TOO!*"

"Do you have the flash drive for the display?"

"Where's Bubba?"

"In this stinky old gym bag."

"What kind of snacks did you bring?"

"I don't see Gibby."

"\He's loading stuff from the shop."

"Are we stopping at McDonald's in Springfield for breakfast?" Someone asked.

"Ha ha," replied Gibby.

"Guess that means no."

"Who's the girls' chaperone?"

"Mrs. Canon. She's cool."

Although an English teacher, Mrs. Cannon liked to work with Academic Decathlon and National Honor Society, and she was

98

curious about robotics competition. Since we'd spend Friday night at a motel in Ft. Smith, the school paid for a woman teacher chaperone and, Mrs. C was a good choice.

Broghan sat by me on the bus. The five hour ride to Ft. Smith could have been longer and I'd have been just fine. We talked and read, and basically just enjoyed being together. He was more mature than most of the other high school guys, maybe because he'd had so much family responsibility with babysitting.

Broghan was friendly to everyone in his quiet way. I was so glad Gibby let him back on the team. Hands down, Broghan was the best spotter and Mr. Gibson recognized that as more important than the stupid duct tape blocking crap argument.

The bus was on Interstate Forty-Four past Springfield and almost to Oklahoma when we exited and headed straight south toward Ft. Smith. My nerves fluttered when we made the turn. I knew we didn't have the resources of the big schools, but it seemed to me that our table display was respectable. Still, Regionals was so much more intense than the Hub where there were only ten schools from two states. Twenty or more championship teams from eight states would converge on Ft. Smith for Regionals.

I didn't want anyone in my group to be embarrassed if we didn't get a decent score in any of the corporation parts. My thoughts whirled. *Better get in one more round of practice.*

"Hey Sara, Shy, and everyone else on the oral presentation team come on up here! Trade seats or scoot someone over. Let's practice again," I yelled.

Metro booth

Circle High School booth

Pit area at Regionals

THIRTY-ONE

+

PAUL

The Friday morning of Regionals, people got to school as usual—bus, pickup, or car and we took our luggage to Gibby's room. We'd stay that night in a cheap-o motel in Ft. Smith but since we couldn't afford a private bus, we had to wait for all the school buses to arrive, dump their loads, and for the one taking us to Ft. Smith to get cleaned and refueled.

My parents would drive down later, not as chaperones, but to cheer for us. I'd room with Phillip and a couple other robot guys. I mean, who wants to stay with your parents on a school trip? Besides, we guys needed to talk strategy for the Saturday competition.

There had been an article in the local paper, *The Lebanon Daily Record*, about our team, and also Lebanon's team was going to robotics regionals, which was cool. Still, most of the Conway community didn't know much about what we were doing. Not like if it had been a sports teams going to regionals. Robotics competition just wasn't a tradition.

Those of us in the Robotics Club knew this was a great thing—getting to regionals with our robot and all. We had our families' support and at least *some* of our teachers had wished us good luck.

When Gibby yelled that it was time to get to the bus, Phil and I double-checked and triple-checked to be sure we had all our stuff. Ft. Smith was much further than Harrison and we'd need to have it all together to practice in the field house in that afternoon. There would be no option for someone to go to school and get anything we might have forgotten, and make it to Ft. Smith in time for practice that evening.

Phillip was grumpy. He and Ceira would miss being in a Christmas parade on Saturday, so the band director was mad at them. I hoped they could get that smoothed over.

Some of the CHS teachers supported our robotics activities; others, not so much. That was frustrating. Some of them would lower a grade if you missed class for robotics.

Grant would lose his starting position for varsity basketball for a week due to missing a tournament game because of Regionals. He'd talked with the coach and said both he and the coach were okay with the decision because it was a team rule. Grant could start again when his bench week was up.

Phillip and I felt good about Fred's changes. I'd analyzed Fred's performance at the Hub competition and improved him as much as I could. In addition to new wheels, he now had a claw that grabbed our hopper and the benzene tanker better than before.

When we pulled into Ft. Smith, Gibby stood at the front of the bus and went over our schedule. Ft. Smith was a juggling act between buildings. Friday afternoon, Ceira's people would be at a building on one side of town and we'd be on the other side, and our motel was somewhere else.

The bus dropped off the corporation team first and everyone wished them good luck. Next, the driver-spotter team and Gibby were dropped at the field house for practice, and Mrs. Cannon went with the rest to the motel to get checked in.

I was anxious to try the Ft. Smith playing field. The field *always* had some quirks. We needed to see how Fred's new wheels maneuvered and find any flaws in the carpet that might trip him up. We grumbled among ourselves about the unfairness of Wichita Homeschool getting to build the playing field *and practice on it* before the competition.

We set up our pit area while Gibby registered. Then the judges measured, weighed, and examined Fred for illegal parts. He passed.

We took our turn on the game field. The field had a lump in it that made Fred bobble so each driver practiced at avoiding that place. In competition it would be nerve-wracking to work around that bump while also trying to remember all the chemicals we had to collect, watch the clock *and* our rivals' robots. They'd be stealing balls, cans, and globes here much more proficiently than any robot at the Hub. Robots would be in and out of all four game field quadrants, including ours.

We watched our competition's practice rounds. There were some pretty slick robots there. We'd hoped for an advance sneak peak but didn't get it. Mr. Gibson sent Mr. Coyle on reconnaissance to Wichita one day when they had some sort of exhibition of robots. Mr. Coyle was supposed to video the action so we'd know how their robot moved. Only thing, Mr. C wasn't much of a tech person and the video was totally dark, so we saw the Wichita robot for the first time when we set up our pit area. We'd teased Mr. Coyle about it and he smiled.

The Wichita robot was really sophisticated. It looked professional. A lot of aerospace engineers sponsored that school.

Soon as Gibby finished registration, he joined us to watch the competition. We talked about the rivals' robots. There were several professional quality robots that didn't look like they came out of high school and even though we expected it, actually seeing them annoyed me. Same ole, same ole.

Gibby told us to keep our cool and reminded us that Fred could maneuver better than some of the other robots because he was streamlined. No frills, all business, our robot.

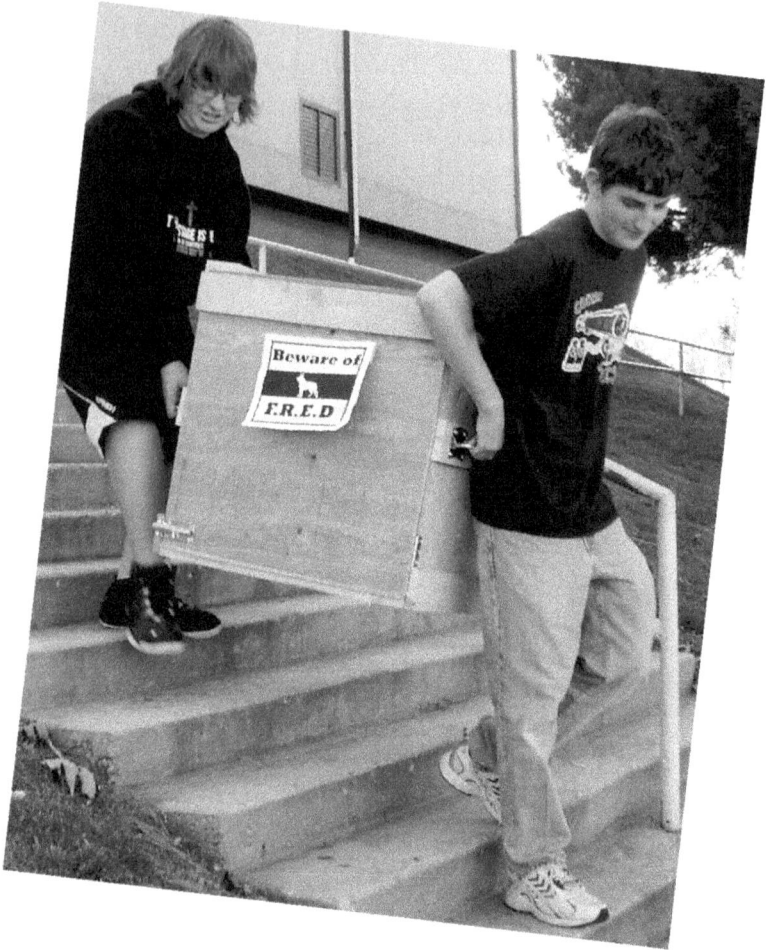

THIRTY-TWO

+

CEIRA

When the bus driver let my group off at the Convention Center, Broghan and the rest wished us good luck. His words and smile gave me confidence.

Regional competition was nerve-wracking. Everything was bigger by tenfold and private schools made up most of the competition. We were painfully aware of our small, rural school status but were determined to hold our own, heads high.

Metro Homeschool from Kansas City people came by to chat. They were always friendly and fun, not like some of the other private schools. Metro got it—that robotics was supposed to be a fun competition, not win or die attitude. You showed good sportsmanship by shaking hands with the losers as well as the winners.

Circle School people were professional, too. We always wished each other good luck.

We set up our table display, then hung around with the other teams and checked out each others' displays while waiting our turn for the oral presentation. We recognized several people from years before. Ah man, some of the table displays were really amazing. The posters had been professionally printed. Money.

Our posters were hand printed.

At 4 p.m. we did our oral presentation. We remembered all our lines and the PowerPoint went smooth, but despite more practice, everyone showed more nerves than at the Hub. Our voices would sort of vibrate at odd times, and my hands were sweaty. I found myself wanting to clutch them in front of me, but that wasn't professional looking so I clasped them behind my back. When we finished, I told Sara that I hoped I didn't look like Mr. Coyle. My tummy is pretty flat but still—.

She laughed and said I looked cool and calm.

While we were doing the corporation stuff, those who weren't on Team Fred as drivers or spotters got to go to the motel. It was El Cheap-O, as usual, but hey, no parents on this trip so… enough said. In the morning, we'd get a free breakfast with cardboard sausage and re-constituted scrambled eggs. Faux food.

The rich schools stayed in the big hotel that attached to the Convention Center. They had an indoor pool which was totally unfair. But that's life.

After our presentation, we packed up our stuff. We'd get our scores Saturday after the semi-final rounds of robot competition. We chit chatted with other teams while we waited for our bus.

After the bus picked us up, we collected Gibby and the robot boys and headed for a supper buffet. Someone joked that it was a food trough. Whatever. It was close to 8 p.m and we were starved. It had been a long day with no lunch, just snacks. We chowed down big time.

You'd think everyone would go to their rooms and crash after supper but most of us watched movies or played games, too pumped to sleep. Broghan and some others came to my room. Broghan and I behaved, although Mr. Gibson was paranoid. He should have been following Shane around, but *nooo*, we got the benefit of his attention. He should know me better than that!

There were five or six of us there for a while, and then most of us crashed in assigned rooms like we were supposed to.

Some kids ran up and down the halls all night—well at least until 3 or 4 a.m. Must have been some freshman, who'd never been out of the county before that trip, poor sheltered things.

I wanted to kill them.

Mr. G. and Shane arguing while group watches

Benzene tanker

Grant, Lloyd, Shane & Mr. Gibson

THIRTY-THREE

+

PAUL

After we practiced with Fred on the playing field Friday afternoon, we had to go to long-winded sessions with engineer-judges who talked about the rules. Wah wah wah. My stomach growled. Phillip grinned and I elbowed him.

Finally, the bus picked us up and we went to a buffet and chowed down. There was a banquet and mixer on campus for robotics people that we went to once and thought was dumb. Besides that, the banquet prices were expensive. Plus after you ate the skimpy meal, there were boring speeches. Our way was better, cheaper, with more food and way more fun.

Back at the motel, a bunch of us hung out. We talked, watched a movie, and I went to bed before midnight. The alarm would ring at 4:30 a.m.

Later, I heard about some dumb stuff some of the others did. Something about buying a butane lighter at a gas station and almost blowing up a car. As long as they were awake enough to drive the robot, I didn't care. They could blow themselves up, just don't mess with Fred!

Saturday morning we ate the motel breakfast and were at the convention center before six a.m., but we weren't the first ones there! We staked out some seats for Conway where Gibby would be close enough with the laptop to see and hear the judges, and relay information to the robot drivers.

We set up our tools in the pit area, plugged the battery charger in, and got Fred ready to make High Octane fuel. Shane tried to fiddle with Fred and I yelled at him. Gibby made him back off.

Everyone was serious during regionals but, most everyone was friendly at the start of the day—even Wichita's team.

Some of the other teams had girl drivers and I thought about Ceira. She could have been a good driver, given the chance. Maybe next year.

The crowd in the field house grew and grew and got louder and louder. At 7:30 a.m. the usual nameless rock music blasts pounded our ears.

Our spirit team arrived right then, and found their seats. They weren't as noisy as they'd been at the Hub. Or maybe it was just that the other schools' spirit teams were so much bigger. And gaudier. Lots of mad scientist-types bounced around in wild wigs and costumes. We only had about twenty-five to thirty people in our spirit team.. But they looked good in their matching High Octane t-shirts.

We saw some people we knew from prior years and talked about our robots and how we practiced. Metro Homeschool in Kansas City had a big building donated to them just for robotics. I couldn't help but be a little jealous. They had a huge support group that filled most of one entire section in the field house, from the ground floor to the top walkway. Several schools had similar spirit teams and bands.

For the past few years, either Wichita Homeschool, Metro Homeschool from Kansas City, or Circle School from Towanda, Kansas won the robot award and the BEST award. I wanted to bump one of them out of the top three slots.

Pep bands were set up on the mezzanine behind each section of seats in the field house, and several schools brought their cheerleaders.

Our group looked so small.

At 8:00 a.m., the announcer took the microphone and did the introduction of schools as robot drivers and spotters paraded in sort of like in the Olympics. I carried Fred very carefully, because if you smooshed his wheels too close to your midsection, they could go off track. Not gonna let that happen! People cheered like crazy while the parade took place. Then we went to our pit area and the first eight robots got ready to rumble.

And all of a sudden, the scoreboard blared and the elimination rounds for High Octane Regionals were underway!

There were two identical playing fields set up, so the floor action was really busy with eight robots going at once, with two sets of judges and referees walking around, monitoring the action.

The central carousel dumped out chemicals, piles of tennis balls, tomato paste cans, and racquet balls. We had a good opening round, no flubs this time, and we did respectably well. The task in elimination rounds was again just to show your robot could gather the different items, not yet make high octane fuel. We collected tons of balls and cans and qualified for semifinals.

One hurdle down.

Yesss! I felt really good about Fred and each of our drivers and spotters did his job well. From our pit area, we watched those who were eliminated dismantle their robots and return the electronics to the officials.

The morning passed quickly.

At lunch break, my ears were ringing from the crowd noise. Since the lines were long at the concession stand, someone had gone for pizza and the spirit team people gobbled down like pigs while Phillip and I tuned up Fred. Gibby then watched Fred so we could go eat.

When we went to get our pizza it was almost gone. We each got one slice and shared the last piece. It wasn't enough. I could have eaten a large one all by myself.

After lunch, we stayed strong and held our own in the ever more intense competition. Our scores alternated between third and fourth place. First place would go to Wichita or Metro, unless something drastic changed in the afternoon. Their robots were amazing, and really sophisticated. Professional.

In the latter part of semifinals, our score slipped and we started using the duct tape ball blocking strategy. Broghan went to sit in the stands. The other teams were surprised and not happy when the balls blocked their robots. You should have seen the look on the Wichita driver's face when we blocked his robot from dumping his load. He was ready to kill! All the Wichita people were furious. They challenged us. An official blew a whistle and all robot action stopped.

The judges went into a huddle.

Uh oh. Would they disqualify us? Kick us out of the competition? Our team gathered at our pit area to see what was going to happen.

Gibby flew out of the stands with his notebook in hand. He flipped to a page as he galloped across the floor to the judges' huddle. He talked, thumped the notebook, talked more. The judges asked questions, looked at his page. They huddled for an excruciatingly long time. Eventually, they ruled the blocking strategy was legal.

Whew!

The Wichita people weren't even the slightest bit friendly to us after that.

Later, one of the people from Circle School told us they'd talked about using a blocking strategy, too. I'm not sure why they didn't. I'm sure they wished they had because we moved up in the scores.

We ended semifinals a solid third place behind Wichita and Metro!

We'd qualified for the final rounds where the top four robots would do battle for the regional championship! We were one of the top four robots from eight states! Fred, who was built with many starts and stops, drove me crazy figuring out the arm, and then almost didn't get his brain programed. Wild.

As at the Hub, it took longer to finish semifinal rounds than planned. It would be a late night ride back to Conway, but getting into finals at Regionals was worth it.

We took a snack break. Ceira had saved us some granola bars and stuff. Yum! Everyone pounded us on the back or said, "way to go," and that kind of thing. We talked with some of our friends from Metro and Circle, then all the drivers went back to the pit area.

Finals brought out the best in our team.

We got more excited each round, knowing we were holding our own with the big schools. Shane got more daredevil each time he took a turn as driver—but in a good way. He'd move Fred toward other robots' territory, then dart in and grab stuff out from under them. Sometimes he moved Fred fast toward another zone like he was going to barge in, and then he'd quickly zoom back into home territory.

Because Shane was so aggressive, no one came into our zone because they knew he'd tag them and they'd be frozen *in our zone* for thirty seconds.

Each team grabbed the right chemicals and made high octane. The benzene tankers worked for every robot. Spotters were right on.

Finals was going to make a grey-haired old man out of me.

Even though the scoreboard was blank during the final rounds, we knew the scores were close. You could feel the intensity on the floor even though you didn't hear the crowds if you were the driver or spotter because of your focus. When taking a turn off the playing field, the crowd noise was huge with people shouting, stomping, blowing noisemakers, pep bands blaring. It was deafening. Exciting!

Object Reminder Card

Overhead shot of robot gaming field

THIRTY-FOUR

+

CEIRA

Holy crap. Bigger, louder, and more dramatic than last year!

Saturday morning, as we settled into the stands in the University of Arkansas-Ft. Smith field house and dug out our home made noisemakers, I couldn't help but envy the snazzy stuff other teams had. They had people in wild wigs—green, blue, white, red—you name it. I saw pom pons, face paint, cheerleaders, pep bands and huge, well-choreographed spirit teams. But we looked good in our High Octane t-shirts and we had Fred, the superstar of Northark College Hub.

Thirty-eight schools from eight states were introduced to start the 8:00 a.m. elimination rounds. The announcer said more than nine-hundred students were there.

He welcomed visitors from the university and the city. There were at least nine hundred more people there just to cheer for their teams.

The announcer explained the High Octane game just a little.

Of the thirty-eight robots starting the day, eight would go into semifinals and by late afternoon, only four would compete in the final rounds.

Next came the parade of schools and as each got announced the big schools had thunderous cheers, stomps, pep band blares and whistles as "their" robot and driver-spotter team paraded into the field house and over to their pit areas.

We screamed and shook our noisemakers when Conway was announced, and chanted, "Go Fred! Go Fred!"

Eight robots took the field and the elimination rounds began.

Gibby gave me the corporation score sheets during a round when Fred wasn't in action. He didn't say anything, but I could tell from his face our scores weren't as good as we'd hoped.

Shyanne, Ashley, Sarah gathered around and we read the judges' notes. The judges criticized us for not having more narrative in the notebook. *Did they want an effin' term paper on each component? We answered each of the questions. We're not professional writers, we're high school students*! Major disappointment. The private schools' stuff looked like it'd been to professional printers for binding. We didn't have money for that! So we weren't in competition for the BEST award. I tried to put the corporation scores out of my mind and focus on the robot action. Fred and the boys were doing well.

Just before lunch, Team Fred moved into semi-final rounds. YEAAAAA!

Broghan was the best spotter. His hand signals were good and he moved fast and sure. He was also fast and accurate when he moved Fred's commodities into the correct zones to make high octane or benzene. I was proud of our team.

Then something happened that made me heartsick.

At lunch break, Sara told Broghan something about me and he wouldn't talk to me. That little– *Ooh! That was so –so seventh grade!* She wouldn't tell me what she'd said and neither would he. He just said he needed to think and for me to let him be for a while. I felt like someone threw a bucket of ice in my face. I texted him but he didn't reply.

I tried to focus on the game, but couldn't shake my feelings of betrayal and frustration. I didn't feel like eating lunch. I texted Broghan, *talk 2 me. Can't xplain if I don't know what Sara said.*

I waited. The robot action went on. Broghan didn't reply. His sister Ashley wasn't speaking to me any more either. She blamed me for Broghan not helping with baby sitting much since we'd been hanging out at my house a lot.

After lunch, there was a rumble going through the crowd as some teams griped about being blocked by our duct tape balls. Everyone hated that Wichita built the playing field and that wasn't fair, but this was something new. Even though the official BEST rules challenged teams to use blocking, when they hadn't thought of it first, they got pissed off.

Semi-finals went on after lunch then all of a sudden it was time for the final rounds. I had been distracted from the action by Broghan's behavior. He sat in the stands away from me when the guys started using the blocking strategy and he still wouldn't reply to my texts.

Conway's Team Fred made it to finals! We whooped and rattled our noisemakers. My stomach growled and I dug out a granola bar. We'd be here 'til late in the evening but that was good because we made it to finals. We wanted to take home a trophy.

The announcer called for a fifteen- minute break while teams that didn't qualify for final rounds dismantled their robots and returned the electronics to the BEST people.

The scoreboard reset to 0000. Final rounds would begin in ten minutes. Rock music assaulted the air and everyone's ears.

We were pumped. We shouted, "Go Fred! Go Fred!" This time, not like the Hub, it really was a mystery as to who would be the first place team.

I couldn't believe my eyes, but Shane was the best driver in the whole competition! He'd drive Fred toward other robots to challenge and taunt them. The other guys would back off and Fred would grab commodities and roar back to our corner.

The duct tape balls got tossed toward Wichita's quadrant and their robot got stuck for precious seconds so they didn't build up much score that round. *YESSSss!* The round ended and we rotated to another corner of the field with three different schools. We racked up more scores, but weren't going to take the lead as easily as we did at the Hub. In fact, at this point, no one knew who was in the lead. I'd catch myself looking at the scoreboard and seeing 0000.

Broghan still wasn't answering my texts. I felt sick.

At the end of one round, there was a huddle of judges. *Now what?* After a few long minutes, the announcer said, "Ladies and gentlemen. We had a timer error. Because of that error, Wichita Homeschool will be allowed to repeat the last round. This was not their fault; it was a timer error."

What? Yeah, right! We didn't boo out loud but we sure wanted to! *Grrr!* Worse, no one else was allowed on the field while Wichita did their solo grab of chemicals and racked up points!

Playing resumed. When Metro's robot wasn't up and ours was, several in their spirit team came over in a group to cheer for Fred. They were *such* neat people! I think partly it was because they were a fellow Missouri team, but mostly, they were just good people. I like them more each time we saw them.

Shane

THIRTY-FIVE

+

PAUL

With each rotation, each of our drivers and spotters did his job with confidence. Lloyd, Grant, Shane, Phillip, and me—we all did good. We racked up points, but so did other teams. Then the judges huddled and we had no idea why.

The announcer took the microphone and said there had been a timer error, and Wichita Homeschool would play a round with no competition.

Timer error? Bull! We were all fit to kill. Disgusted. How to define "fair?" Not that! The other schools just seemed to shrug it off. I looked toward the stands and could see Gibby shaking his head. I knew he would later complain to headquarters, but I didn't think it would do any good. Still, I was glad to know he'd at least have a complaint on record. The same thing had happened the year before.

We continued the duct ball blocking strategy. Other teams didn't like it but I think that was because they wished they'd thought of it first. I kinda hated that Broghan wasn't with us, though. He went to the stands when we started it. At the end of each round, we nabbed those balls so we could reuse them. No one smashed any of them, but I'll bet they thought about it.

After Wichita's replayed round, we knew they'd gone up in scores because they collected chemicals without anyone out there to block them or beat them to the most important items. Bummer! I hoped that other schools would lodge formal complaints.

We went back to the game field and our guys really worked to collect chemicals and the benzene tanker. I hoped that we'd end up with a trophy. I wanted a trophy so bad I could taste it.

It caught me by surprise when the final buzzer rang. Everyone automatically froze. You had to immediately stop moving the robot, and the spotter couldn't move any more commodities into the score zones.

I took a breath and looked around. Crowd noise was loud but rock music from the field house system blasted even louder. Everyone put their robots into their pit areas then we all went to sit with our schools in the stands.

Judges tallied the final scores for several long, agonizing minutes.

Finally, the announcer took the mic and the rock music stopped. First, they announced the corporation awards. For the overall BEST award, Wichita Homeschool took first, Oklahoma City Homeschool came in second, Metro Homeschool took third. No surprise. I glanced at Ceira and smiled. She shrugged. Her face looked really sad.

During award announcements, we applauded for everyone, and especially so for Metro Homeschool. Photographers took pictures of each team with their trophy. The BEST award trophy was huge, even bigger than the one for the robot winner.

My hands were sweaty as I waited for the robot scores. The announcer dragged it out, more dramatic with each announcement. I just wanted him to blurt it out.

Wichita's robot took first place, no surprise. I'll be they wouldn't have without that free round. Kittson Central High School from Hallock, Minnesota took second place, Ambassadors for Christ Academy from Bentonville, Arkansas took third and we, little ole Conway, Missouri took fourth place against the hottest robots in eight states. Not bad. Not bad at all. Phillip and I high-fived.

The end of award announcements and the photo op left me with mixed feelings. Our corporation didn't do well but Gibby said we did very well against the mega bucks competitors. I was proud of Fred, and so was Phillip and the rest of the team.

We had tons of photo ops, including newspaper people from Ft. Smith. Several television stations had camera crews there. I hoped the coverage would make it to our Springfield TV stations. But even so, unless you were actually in that field house all day, you just didn't get the excitement of robotics competition. It was better than basketball, baseball, and track rolled into one.

THIRTY-SIX

+

CEIRA

As we watched the teams get ready for award announcements, I saw Shane flirting with a girl from Wichita. *Wha-?* I never knew if he was serious or just trying to piss off her brother, who was one of their robot drivers.

The web page first place went to a school from Fargo, North Dakota. The t-shirt design award went to a school in Minnesota. The most photogenic robot, the beauty contest, went to Shattuck, Oklahoma. The most elegant machine first place went to Wichita Homeschool, and Conway took second!

Another prize for engineer Paul. Yesss! The Founders award went to Conway. I'm not sure what that's all about but it was another trophy so awesome! The BEST award for the corporation trophy went to—ugh—Wichita Homeschool.

Wichita Homeschool also took first in the robot competition. That was the second straight year they finished in the top spot. Without the solo replay of that one round, they wouldn't have won High Octane.

The photo op was big with lots of newspaper people with cameras and TV crews. Some schools had SKYPEd play-by-play to their home people throughout the competition. Cool.

Our guys did good, damned good, but Fred finished fourth. Still, that would be a nifty trophy to add to the display in Gibby's room. I think our guys were pleased. I mean, third at regionals out

119

of thirty-eight schools with bigger budgets, and huge support from megabucks aerospace plants and other ginormous businesses that you don't find in rural Missouri.

Broghan still wasn't answering my texts, so I had to guess by his face that he was okay with fourth place. He wouldn't make eye contact.

Just as people started gathering up stuff to head for the buses, the announcer again took the microphone. It was almost 8 p.m. and the event was supposed to end at 5 p.m.

"Ladies and gentlemen, we again congratulate all the winners tonight, and that includes every school that brought a robot here. You're all winners!" He waited for crowd cheers to stop.

He drew out each word as he said, "And now, for the first time ever, the BEST organization is sponsoring a national championship. The top three schools' robots *and* the top three schools with the BEST award are invited from each region all over the United States. That is going to be something, let me tell you. Whoo Boy! And because the top robot at our Regional competition here in Ft. Worth also won the BEST award, we're allowed to invite the next place robot to go to Dallas." The announcer made a dramatic pause, "and that - is - Con-way, Missouri! You're invited to nationals!"

We looked at each other in shock. The announcer continued, "The third weekend in April, winning robots from every regional competition in the U.S. will be invited to the convention center in Dallas, Texas. We'll share the floor with the VEX Robotics college convention. SEE – YOU - THERE!"

Rock music and feet thundering out of the stands blended with crowd cheers. I looked for Gibby's reaction but his back was to me while he spoke to Paul and Phil.

On the bus ride home, the robot boys and Gibby talked about nationals. I didn't so much, because our club had no more money. Besides, if I tried to take the table display and other corporation stuff, I'd need professional help with spiffing it up and who would do that?

My phone beeped. Oh!! Broghan finally answered my text messages. We found a quiet seat toward the back of the bus to talk, kissed and talk more. I was so happy!

It was way past midnight when we finally pulled into Conway. Some of the parents griped because we were so late but most were just happy to see us home and thrilled to see the trophies. I'd left my car at the school so I could drive myself home, and I'd called my dad to tell him about regionals and the invitation to nationals. He and my stepmom were really happy for us.

Paul, Phil, and the other robot guys were super excited. I heard Gibby tell them, "Guys you've got a very good robot. If you rebuild the weak parts and practice until you've got it perfect, you can make a respectable showing at nationals."

So I guess Gibby figured we had some hidden money somewhere to take the team to nationals. Awesome.

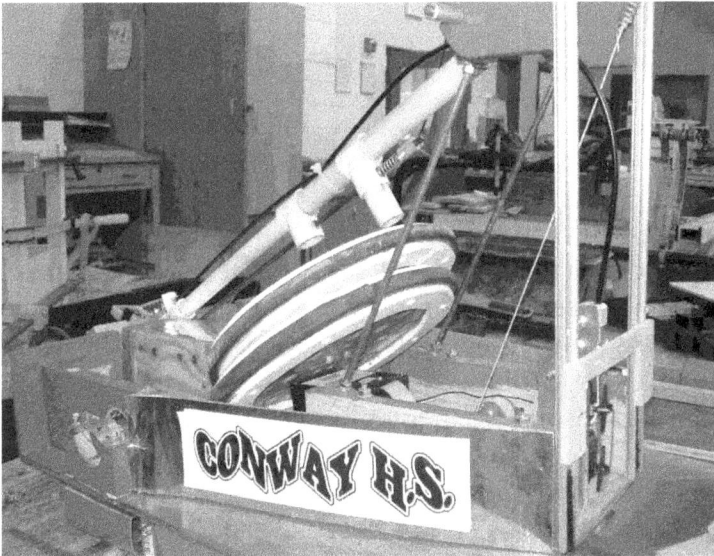

Worn out robot

THIRTY-SEVEN

✦

CEIRA

Monday afternoon after school, only a few robotics people gathered in Gibby's room to talk about nationals. Paul, Phil, Broghan, and I were there along with a couple of excited sophomore robot apprentices. None of the three seniors showed.

Gibby told us what he found out from the superintendent. The club had only a few dollars left. Worse, he had no idea whether Dr. Berger could, or would give us excused absences from school. He had to talk with Mr. Lowrance, our principal, and also to his state department advisors. Dr. Berger didn't know if it was legal for a public school to approve, much less fund a team of students and teachers for such an event. He had to have some sort of legal backup before he could give us the go-ahead.

As for excused absences, Mr. Lowrance had never been a fan of Robotics Club and we didn't expect his attitude to change. We grumbled loudly amongst ourselves.

Mr. Coyle showed up at the meeting long enough to say, "Let me work on raising money. I can tap some alumni."

Mr. Gibson said, "Keep me posted on that, please."

Mr. Coyle nodded.

That perked us up. We looked at each other and went back to talking. That is, Gibby talked. He kind of blurted, "I'm really sorry, but the corporation stuff just isn't up to par. I mean you guys saw the other schools' stuff. No way can we compete with that."

He didn't even make eye contact with me. I about choked! I said, "I know. But it could be if we had some money for it, too. We can make it better."

Mr. Gibson said, "Mr. Coyle didn't say anything about finding more money for the table display. He was talking about the robot and we're not required to take the corporation to nationals."

Paul gave me a look of sympathy.

I didn't say anything more.

Paul said he and Phillip would need to rebuild Fred because all the moving parts were worn out. He'd learned some things that needed to be made better and it was okay to do that without breaking any BEST rules. In fact, BEST engineers wanted teams to keep improving their robots when they went from one competition to the next. Paul said something about change to make it even lighter, stronger and faster.

Mr. Gibson said, "You guys all drove Fred well. If you can get your speed up where each one of you is driving even better, you can make a good showing," He paused, " *If* we get permission to go."

Christmas break was coming up and Paul said he'd make a design for a new, improved Fred then. Gibby said he'd talk with the principal and work on getting permission to go to nationals.

Nothing more was said about the corporation part.

Broghan and I left the meeting after that. He said he didn't care about going to nationals if I didn't, and put his arm around me. That helped soothe my feelings but I was still disappointed. I could make the display better, but I also knew when to fight a battle and when the battle was over before it began. I decided to let the corporation thing go while Gibson worked on getting the robot boys to Dallas. I assumed he was talking about the whole team.

After Christmas break I'd be getting ready for the ACT, band spring contests, and prom. BEST nationals would be the same week as my junior prom. It might not be important to the guys, but missing prom would have been a big thing for me. And no, I wasn't compensating for being told I couldn't go to Dallas. I could have pushed for it and I think my dad and grandma would have made sure I had the money, but there just wasn't enough time or help to make our corporation stuff into the professional caliber marketing package it needed to be for the national stage.

And then I had another battle to fight.

The band teacher tried to lower my grade from "A" to "B" because I'd missed a parade for Regionals but I argued my point and got my "A" reinstated. No small factor was my being selected for honors band and I was taking a trombone solo to the spring regional music contest.

Homework, band, and prom prep took my extra-curricular time during January, February and March, instead of the robotics corporation stuff. But it didn't totally take all my heart. I did want to go to Dallas, but I wasn't about to let Gibby know that.

Dr. Berger & Mr. Coyle

Worried Mr. Gibson

THIRTY-EIGHT

+

PAUL

Our invitation to nationals earned us a decent sized article in the local newspaper and front page headlines in *The Chronicle*, an insert in *The Lebanon Daily Record* newspaper that went to the entire Conway community. After that, we heard lots of congratulations when we went into a gas station or our local grocery store, and from *some* teachers at school.

Getting permission from the school to go to the first ever BEST national championship should have been no more difficult than getting permission for a bus to take the basketball team, the cheerleaders, and the pep squad to a tournament. Instead, it became a huge battle. Our principal was a sports nut and an academic thing like robotics just didn't get his jets going.

Never before in its history had Conway High School in our Laclede County rural school district had a chance to do anything academic at the national level. You'd think that would have the entire community behind us donating money and lobbying the superintendent, but it just wasn't in their habit to support robotics like they did athletics.

Even after the media coverage that we were invited to compete at national robotics competition the "discussion" went on. And on and on.

Dr. Berger said his superintendent friend in Lebanon told a wrestling team they couldn't do a national trip to a competition, so that was a bummer. He'd hoped for support, a precedent from Lebanon. He said he'd call the Jeff City people again because they were the ultimate authority and he'd talk to someone who dealt with school law.

The Old Man got nervous.

Phillip and I grew discouraged. It seemed the whole state focused on sports. Who did they think was going to build the machines that saved their lives in medicine or their new cars and enormous pickups? The robot boys, that's who! The lack of support was more disappointing than finding the shop equipment in worse shape each day.

Ceira got so angry she gave up trying to do anything about the corporation component and threw what energy she could into helping us guys get Fred to nationals. Mostly, that meant mouthing off but we were among the smartest students in school, and several teachers rallied behind us.

We got some donations and vocal support from the few businesses and some clubs in Lebanon like Rotary or something where Dr. Berger was a member. The Conway Flywheel Club that dad and I belonged to gave money, and more alumni sent money, but I had no idea how much all that added up to.

One day in the hall at school, Mr. Coyle gave me a check for a thousand dollars for the club. He didn't say where the money came from. That perked me up quite a bit but it made me super nervous to carry that check around until I had a chance to take it to the office. The secretary smiled and said they would deposit it into our robotics account.

After one heated discussion with the principal, Mr. Gibson went to Dr. Berger. He said at the very least, he wanted to take the two robot chief designer-builders, Phillip and me, to nationals. He'd drive his own vehicle if he had to. I don't know what else he said, but it helped!

Dr. Berger said his government guy hadn't found anything prohibiting a public school to take students on an academic trip such as that, so by default, that meant it was okay to go. Dr. Berger also found a few hundred dollars in some fund that he said would pay for the school van, but Mr. Gibson would have to drive it. Gibby had no problem with that.

Mr. Coyle had raised enough funds to pay for food and a couple of motel rooms, one for Gibby and one for Phillip and me.

We were going to take Fred to Dallas!

When I asked about the rest of the team, Gibby said there wouldn't be any room in the van for more than the two robot engineers, Fred, the tool box and spare parts. So he'd argued for taking only Phillip and me to Dallas.

I didn't know what to think about that.

Fred's moving parts were totally worn out, but we had time to totally rebuild him in the four months before Nationals. Phillip and I steadily worked on him, and saved plenty of time for practice. We needed to improve our speed and accuracy driving the new Fred.

My parents planned to go to nationals. Phillips's parents wanted to, but didn't have anyone to run the dairy business for them to be gone that long.

Gibby protected us in the rebuilding process by not allowing any of the goof offs into the shop while we worked. Most of the time it was just Phillip and me, anyway, so we made good time. I'd made some design changes to streamline Fred's body to make him lighter. The new wheels Phillip built ran even smoother than before. I did Fred's paint job this time, but used Sara's design because it was good and I'm not creative with that sort of thing.

The shop slackers still did their destructive stuff during the day and one time after school, I got so frustrated with broken equipment that I picked up a hammer and threw it across the shop. I didn't aim at anything, I just let go. It gouged a hole where it hit the floor. Oddly, I felt better.

School, homework, robot rebuilding, and family activities. That was pretty much the world for Phillip and me. We weren't party guys and didn't have parades so robot rebuilding dominated our after-school lives that spring. We saw Ceira in Trig class, but she didn't come to after school robot building very much. She and Broghan were planning prom activities.

When New Fred was ready to roll, Phillip wrote "33 days" on the top of a chalkboard in Gibby's room. That was count down until Dallas.

We practiced after school each day, including the seniors, now that basketball season was over. Gibby became a taskmaster at levels we hadn't seen before. He wouldn't allow anyone to text or even hint at goofing off during practice. "You need to practice until you have perfect practice," he bellowed.

We did. Even I got so fast driving Fred that I didn't hesitate one fraction of a second picking up commodities from the floor or grabbing an overhead globe. The globes now hung by paper clips attached to fishing line suspended overhead. The first of the year, the regulations called for Velcro attachments to the overhead fishing line. That stuff was hard to detach. The paperclip came off smoother, but if you weren't one hundred percent accurate, you could snag the clip and mess up big time.

We each took turns being driver and spotter. Broghan watched some, but he wasn't part of the team any more.

Shane still got on everyone's nerves by being mouthy, but even he got better at driving and spotting. He knew Gibby would kick him off the team for good if he messed up practice time at this point, so he curtailed the worst of his attitude.

After basketball, track and baseball dominated community time and most of the administration's attention that long spring.

Then another battle took place, and this time parents got involved.

When it came time to pack for Dallas, the three seniors didn't get official permission in the form of excused absences from school to go. Grant's mom, Teresa, hit the roof.

I could understand her arguing that it was a team effort that won the invitation to nationals. If just Phillip and me went, we'd have no chance at making a decent game. We'd be lucky to make it out of the elimination rounds.

Mrs. Rumfelt made a big scene with the superintendent and Lloyd's parents did, too, for his sake as well as Grant and Shane. Shane was the best driver, no doubt about that. Phillip and I were good with Fred, but we didn't make up a team and we'd only be at Dallas for a token showing without the others.

To make it more complicated, Gibby said that if Ceira or *any* girls went to Dallas, the school would have to come up with a chaperone, and it would cost a lot more money. And then the rest of the club would want to go, so everything was less complicated this way.

I had no response. Maybe I was selfish. Maybe I thought we

didn't have a chance of making it past semifinals even with the three seniors. Maybe I didn't want to jump into a discussion I couldn't win.

In the end, I was glad the seniors' parents decided the guys were going to Dallas, but sorry they had to pay their own way.

Some of our teachers wished us well on the Thursday morning we left school for nationals; others said nothing. I was glad for the support of those who understood how much this meant for Conway.

When the van was loaded with Fred in his crate, the tool box, spare batteries, Gibby, Phillip, me, and our luggage, there actually was no room for anyone or anything else.

Lloyd's Mom drove their family van and took the seniors, plus Grant's mom to Dallas.

Mr. Coyle drove himself down. My parents, sister, and little brother would drive down on Friday.

It was a *really* long drive from Conway, Missouri to Dallas, Texas. With rest stops, it took seven hours. We got tired and slap happy, but Gibby kept us going. Both Phillip and I get car sick if we try to read or do any electronic games, so we were stuck with conversation, watching the changing landscape, and making fun of billboards and strangely named roadside cafes. Toad Suck Grill really set us off.

Late in the afternoon, skyline of Dallas got our attention. Mr. Gibson had been there before, but neither Phillip or I had. We were impressed. The skyline was awesome, and also intimidating. We stopped at some place that had a bunch of life sized Longhorn steer statues as a tribute to the trail drive history of the region.We took pictures. Phillip joked about getting some of those things on his farm. He joked they would be less messy than live cattle.

We checked into a hotel Mr. Gibson had lined up. We'd be there three nights and return to Conway on Sunday instead of the usual Saturday night return; otherwise, we'd be driving all night. Dallas was a lot farther away than Ft. Smith.

Then we checked out the competition location. The Dallas, Texas Convention Center was the biggest building I'd ever seen in my life.

Phillip and I just hoped we wouldn't embarrass ourselves with Fred.

Dallas

Paul & Phil

Wetumpka High School - Dallas

THIRTY-NINE

✦

CEIRA

"That sucks!" My reaction when I heard the three seniors didn't get official permission to go to Dallas after all their practice getting ready. That was even more unfair than me and my helpers not getting to take the corporation stuff and that *still* burned me.

When Grant's mom pitched hissy to the superintendent, she was right to do so. I know Lloyd's parents spoke up too. Good!

I know, I know, at the first of the year, I didn't think those guys would contribute but I had to admit, they proved their worth.

When the time came to leave for nationals, Teresa Rumfelt, and Lloyd's mom, Ruby, drove the three seniors down to Dallas in the Oberbeck's van. Bad enough they had to take their own van, but they didn't even get friggin' excused absences from school to go to a national *academic* event! I don't know who paid for their gas and the hotel.

Others in Robotics Club were angry too, but what could we do?

And yeah, I would have loved to go, but I still don't know if I would've chosen that trip over my junior prom. My first prom. My first prom date with Broghan. It's just that *I* wanted to be the one to make the choice, not have it forced on me.

Prom was Saturday night, same time as the final rounds of the robotics championship if our guys made it into the final round. I'd watched them practice—all five of them—and they were good!

Each one could make Fred fly like a NASCAR racer.

Funny to think about the year's start, and that Paul predicted the theme would be about speed and a race car. They'd now actually made Fred into a speed demon, streamlining him after each competition. They also their increased driving speed. Our boys would not embarrass themselves at finals, I felt sure of that. They weren't that confident themselves, but I was.

Prom prep took most of Saturday. Our prom theme was "Night With the Stars." We had giant sets of scaffolding, but some of the people wouldn't get up onto it to reach the gymnasium rafters. Wimps. I did and so did Broghan. We spent more than four hours decorating the gym, transforming it into a ballroom by covering the beams with gossamer.

My prom dress was awesome, but I could have worn something from the hand-me-down store and felt fine because I was going with Broghan. We went with a retro gangster Hollywood look for our night as movie stars. My first formal dress was mermaid cut turquoise blue, with a low cut back. Broghan said it made my eyes reflect blue-green.

I put my hair up. I'd bought $12 shoes that were walkable and danceable. Broghan wore a pinstripe tux with a fedora. He was gorgeous!

We were on the dance floor when someone got a call about nationals and the robotics people went into a huddle.

Broghan & Ceira

Parade of schools - Dallas

Grant in Dallas

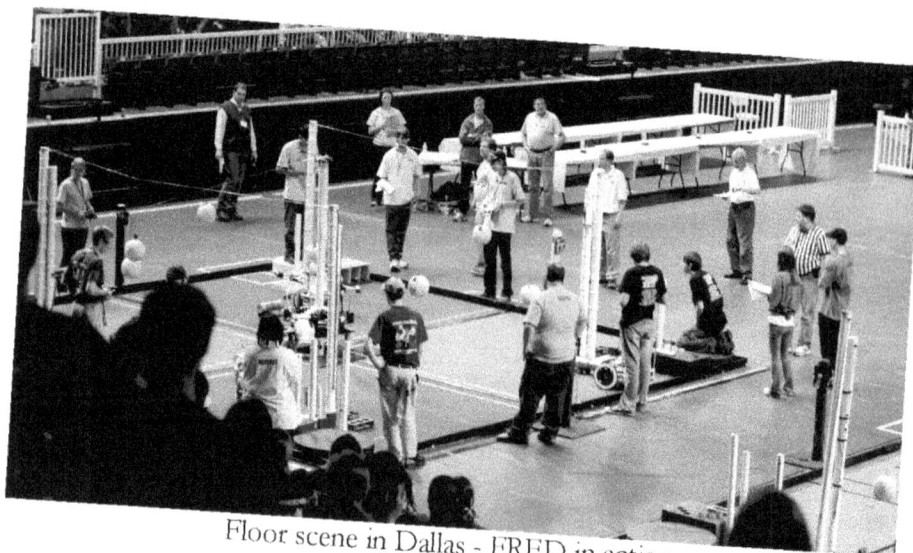

Floor scene in Dallas - FRED in action

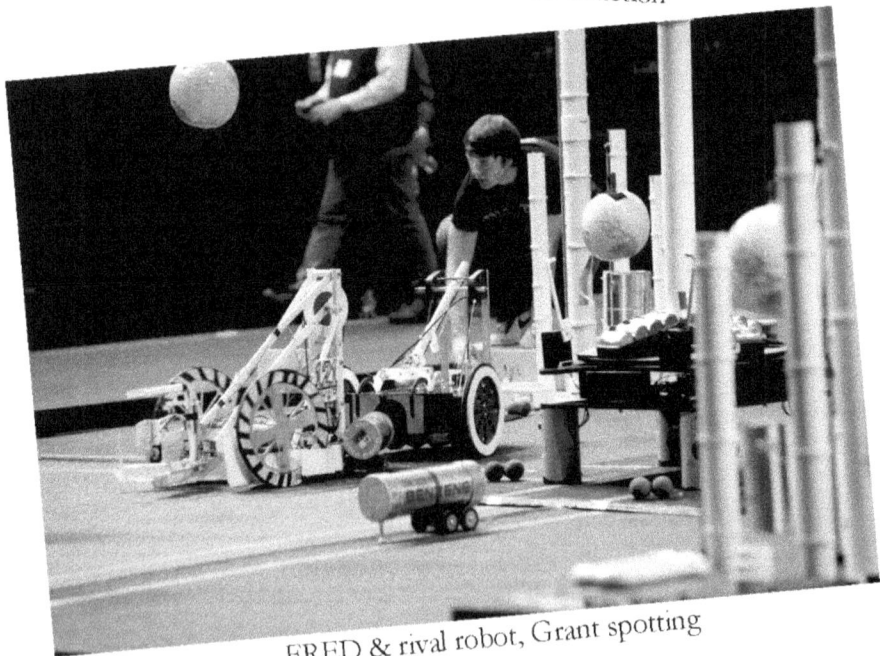

FRED & rival robot, Grant spotting

FRED

What's Gibby pointing at?

Nationals long shot

Paul carrying FRED

Paul closeup, in Dallas

FORTY

PAUL

Shane and Grant explored downtown Dallas Thursday night after the rest of us went to bed. Lloyd stayed in the room. He either wanted to get sleep before the big day or maybe he worried that his Mom would find out if he left.

Shane's one hundred dollars for the weekend was in the form of dollar coins, which he'd put into his pockets. He and Grant gelled their hair up and went strolling the streets after dark. In a huge city where anything could've happened. Jingling money. Maybe they looked so weird that not even thugs went close to them. I don't actually know what they were thinking.

Gibson never found out.

The Friday pre-competition meetings were boring until the Old Man asked the officials for clarification of the rule about blocking. He wanted to make sure we wouldn't get challenged on the floor and made to forfeit a round when we used the duct tape balls. He bellowed out his question from way up in the stands in that big group of teachers, students, and engineers, then loped down to the floor with his 3-ring binder.

After a big huddle, the judges made us cut into one of the duct tape balls to check the construction. We only had six of the things so hated to lose one that way. They said we'd be legal to use them.

After the meetings, we practiced. The two playing fields were well-built, and glory be, smooth! We worked out some of our nerves by doing what we'd practiced until it was second-nature to maneuver a robot around the game floor.

We also scoped out the other robots and teams. There were some awesome machines there, but a few were overbuilt. I like Fred's streamlined look. Our robot was basic, no frills except for the shiny paint that really stood out under the lights.

We ate somewhere that night, but I don't remember where, or what I ate.

Saturday morning we were up early, ate breakfast in the hotel, then all of us went to the Dallas Convention Center.

The parking lot was full of luxury vans and charter buses. Circle High school had their traveling workshop van there. In that van, they had the capability to rebuild any part that might malfunction. We couldn't imagine the money they spent on that thing.

The gigantic convention center was full of people early on. There were huge spirit teams from other high schools, and tons of college people with the Vex robotics stuff.

I doubt that a Super Bowl crowd generated more excitement and noise than the eople inside the Dallas convention center there for high school and college robotics action. The air radiated excitement.

My parents, Grant's mom, Lloyd's mom, and Mr. Coyle made up our cheering section. You had to laugh. From the floor where the playing fields were, the five in our spirit team were lost from view.

Our pit area was up an escalator on the mezzanine level. Not good. We had to carry Fred upstairs after each round. We had no choice but to set up our tools and battery charger there. As with regionals, the team members who weren't taking a turn driving, would keep an eye on our stuff. I was there every minute I wasn't driving to make sure no one on ours or any other team fiddled with Fred or "borrowed" any tools.

Down on the main floor, the BEST championship took up only about a third of the convention center floor, and the VEX college robotics event filled the rest. The college people were interested in all the variety of the high school robots. Funny. Their event required each team to build identical robots. That sounded kinda boring to watch. We used the same raw materials, but each robot looked very different.

Everyone but us in BEST Nationals had a big, fancy booth and table display. We checked out the display booths, talked with people, and Phillip took a bunch of pictures.

R2-D2 buzzed around the floor. Cool. Funny, though, to see how little movement that robot actually had in real life. Fred had way more moves. Still, it was awesome to see the world's most famous robot right there on the floor in front of us.

When I looked into the stands, it was overwhelming to see the enormity of the event. But the usual rock music blared and the game field was the same as the Hub and Regionals, except for the totally smooth floor. That, and the fact that the best robots in the entire United States would compete.

Like regionals, BEST people had two complete fields set up so that eight robot teams at once would compete during the elimination rounds and semifinals. I tried to focus on keeping calm. We wished Metro and Circle good luck. Wichita ignored us.

I knew in my heart we were ready, but my stomach was jittery. Phillip's cheeks were bright pink even before we got going. Everyone on the team was way pumped.

Gibby motioned us to get in line for the parade of schools. When we came into the convention center with all our different robots, the college people watched and cheered!

After the introductions, Gibby pulled our team into a huddle and reminded us the field was just like all the others we'd been on. We each could drive and spot much better than we did at regionals, so all we had to do was stay focused and drive.

Yeah, well, easy for him to say.

And then it was our turn on the field for elimination rounds.

BLAATTTT!! We flew into action. Fred performed flawlessly. During those first rounds, we worked out our jitters, grabbed our share of fake chemicals, and "yeeha!" I started having fun.

We qualified for semifinals.

One hurdle down. I looked at the stands where I thought my parents were just as Gibby came out of the crowd to congratulate us and talk strategy for semifinals. Basically he said, "Keep doing what you're doing."

We started semifinals no longer worried about embarrassing ourselves. I looked to the stands, and saw a sea of people in that giant, noisy convention center. Somewhere in that crowd, I'm sure my sister and little brother waved.

As semifinal rounds went on, we got better and better and the morning flew by. At lunch break, Gibby complimented us and made sure we each had something to eat.

Our parents found us. My parents were having a grand time, especially my dad who's a techno geek. My little sister, Rachel, had skipped school to be there and she scoped out the display booths. All of them were totally in awe of the place and overjoyed to see us doing so well.

Lloyd's mom, Ruby, said something about us representing not only our school but also our state and the entire Midwest. That made me feel the pressure again, but we had already proved we belonged there, in the national competition.

After lunch, we continued our game the same as when we'd practiced over and over and over while Gibby did his drill sergeant-coach thing. It paid off. We were smooth. We collected chemicals, made high octane, carried Fred safely up the escalator to the pit area and back down each time we had a turn on the playing field.

We finished semifinals at the top score position. The top position! I could hardly believe my eyes when I saw the scoreboard!

Gibby climbed out of the stands and congratulated us. Phillip and I focused on Fred and the pit area. We made sure that Fred was fully functioning. We charged batteries, replaced rubber bands, checked motors servos and all connections.

Then things got serious.

The scoreboard went blank. We got ready for the final rounds of competition. Final rounds of Nationals! Gibby told us to just keep doing what we'd practiced and we'd finish with a trophy.

I took a deep breath and focused on the game field. I was nervous again. Phillip's cheeks were pink. Grant's face was intense, intimidating. Lloyd was calm, but his eyes were really alert. Shane was revved to the max.

In the final ten rounds, we again faced our archenemy Wichita Homeschool, and also faced our friends from Kansas City, Metro Homeschool. To be fair, we used the duct ball blocking strategy against everyone.

We knew we'd be taking home a trophy, and figured a third or maybe even second place at Nationals was a possibility.

BZZZZZZAATT! Balls and cans flew out of the central carousel and rolled in all directions onto the game field. Four robots roared into action.

Shane was a wild man. He racked up points—all the chemicals he could grab in a round—then drove Fred through everyone else's field, just to mess with them. He could see the entire four quadrants at once. Somehow, he could tell which moves the drivers were going to make next.

He was amazing round after round. Then he did something none of us could have imagined, not even from Shane. We were in a round where Wichita was next to us in the four-quadrant field. Fred had gathered all possible balls, cans, globe, the benzene tanker, and had nothing to do so Shane kept Fred going, buzzing their robot. Messing with them. You could tell they were angry. Then he deliberately slowed Fred down and let their driver tag him.

The judge's whistle blew and Shane dropped his hands and grinned.

Fred would be frozen for thirty seconds. Only thing, Shane had deliberately blocked Wichita from their docking station!

Their driver tried to make their robot push Fred out of the way, but Fred didn't budge. The driver backed up the Wichita robot, revved the motor, and took off fast. The robot smashed into Fred. *Bam*!

TWEETTT! A judge blew his whistle and shook his head, "No." The rules didn't allow for destructive behavior against a competitor's robot. Fred sat. Shane grinned. We now knew the difference between "nudging" and "destructive behavior" in the BEST rules.

Wichita ended that round with hardly any points.

The afternoon flew by. Each round, we made tons of High Octane fuel. I caught myself looking at the scoreboard only to be reminded it would be blank until the final whistle.

In the stands, some of the teams that didn't make it into finals started cheering for underdog Conway!

Later, I found out word got around that a tiny rural school from Missouri was kicking robot butt on the national field. Ruby Oberbeck, Lloyd's mom, got all teary-eyed and mushy when she told us.

We were three rounds from the end when disaster happened! Fred lost a wheel mount!

After each round, whoever was driving carried Fred up the escalator to the pit area. I'd tried to tell the others to not hug Fed against the belly, but to hold him out and make sure his wheels weren't getting mashed.

I won't say who, but after the seventh round, someone hugged Fred and a wheel mount slipped off. I couldn't get it repaired in time for our next round. I wanted to hurt someone. Instead, I focused on trying to repair Fred.

Phillip went looking for Gibby, who was in the stands as required. He couldn't work on Fred, but needed to know what was going on. We were due on the floor again and had to compete or lose out. We'd come too far, worked too hard to give up now!

Fred limped around the competition floor but we couldn't get many chemicals into our docking area. We were losing ground each round!

The other drivers smelled blood.

I worked on Fred each break. We agonized over the possibility of someone catching up to us. We'd come so far only to have a mechanical malfunction that could have been avoided!

Gibby came to the pit area. He couldn't help me with tools, but he kept the others away while I worked. Having that space helped me calm down.

Finally! I got Fred's wheels working again. We raced back down the escalator to take our final turn. Our final turn and Fred was fully functioning. We'd again look good on the field.

We had no idea about our position in the scores. I feared the worst.

Those two bad rounds had felt like a lifetime. Nothing I'd experienced before or since could compare to that day's emotional roller coaster. Agony!

Fred performed smoothly, and we racked up some more points. We had no idea where we were in comparison to the others.

BZZZZTTT! The competition ended. Judges huddled to calculate the final scores.

Phillip and I didn't know whether to laugh, cry, or pass out. We figured we'd end up last. But still, last at Nationals wasn't anything to be ashamed about.

It seemed like hours before the judges finally had the scores tallied.

The announcer took the mic and the rock music blasts abruptly stopped.

I held my breath.

At the end of that long, crazy-wild day, the unthinkable happened. *No one caught us*! We had the number one robot in the United States! Us – Anonymous High School from Dinkytown, Missouri. Wow! *WOW!*

We cleaned house.

When all the scores were tallied, not only were we the fastest and most accurate team, Fred also received the trophy as the most elegant robot, *and* the trophy for the most robust robot. Holy cow!

When it came time for pictures, we were numb. It shows on our faces in the photos. The enormity of the day's events were too much for immediate comprehension. We were *all* shell-shocked as that huge crowd focused on us. *Cheered* for us.

The national spotlight was overwhelming. While the bulbs flashed and the TV cameras and my mom's camera whirred, my brain flashed back through the long year. I thought of the times I felt like quitting, the times the team members squabbled, the lack of community and school support, the broken school shop equipment—it all faded to nothing while the flashbulbs went off again and again and again and I stared at our trophies.

The official photographer took several formal poses of us with our trophies, then he asked us to goof off for a shot. It's Gibby's fav shot, I think. He's got a fake "grrr" look and is reaching toward Shane, who is mock-cowering. Grant is grinning, Phillip and I have giant smiles and weird poses. Lloyd is standing all quiet, as usual, but smiling.

From the corner of my eye, at the edge of the crowd I could see my little brother jumping up and down. That made me laugh.

When we went to collect our stuff from the pit, a Wichita driver said, "I hate you" and cried. Shane took his picture. Grant laughed.

I wondered how we'd get everything in the school's van because those trophies were huge but we'd manage even if we had to UPS some luggage home. Those trophies were going to ride with us!

That night, Gibby took Phillip and me out for giant, Texas-sized steak dinners. The best steak, ever! Too bad the seniors and Mr. Coyle had already started their trip home.

Sunday morning we did some sight-seeing around Dallas before turning the school's van toward home.

The trip back went much faster than the trip down. The enormity of our accomplishment started to really sink in. Gibby talked about potential scholarship offers for us but I just wanted to relax now, finally, and enjoy what we'd accomplished.

Shell shocked winners

National Trophy - Mr. Gibson, Lloyd, Grant, Shane, Phillip & Paul

Wichita Homeschool Warriors Booth

Sponsors

Paul & Phil - Most Robust Machine

FORTY-ONE

✦

PAUL

On Monday morning, Phillip, Lloyd, and I were back in school. Shane and Grant slept in. At school, most of the teachers treated us like rock stars—for a few minutes, anyway. I think I smiled all day.

Ceira and the other robotics people were ecstatic. At prom the robotics people huddled in a corner, off the dance floor, and cheered when they got the call about our win.

TV crews and newspaper reporters from Springfield, Marshfield, and Lebanon came to school to interview us. We made regional TV news for a couple of days. Ceira was interviewed, too. Awesome.

I think everyone in Conway finally heard about robotics.

Because he missed school without official permission, Lloyd missed a test and the teacher gave him an "F" on it. On his transcript, he made his first ever grade of "B." He was upset, but his parents told him not to sweat it, because having the national robotics championship on his resume would stand out way more than anything else for a long, long time.

No kidding. Out of 10,000 students who started the BEST High Octane year, we five on Team Fred were at the top of the heap.

Sweet victory!

Robotics Trophies

EPILOGUE

Paul Coryell -   In his senior year of high school, he stayed busy with a new robot, and didn't apply to the big name private colleges. Instead, he used his Missouri A+ scholarship funds to pay for two years at the local Community College (Ozarks Technical College). Just after high school graduation, however, his appendix ruptured and he had to take a semester off to recover. At press time, he is finishing two years at OCT and will transfer to Missouri University Science & Technology and major in Mechanical Engineering. For fun, he is restoring a Model T and at press time, he has the car drivable.

Phillip Foust – In his senior year of high school, he, too kept busy with the new robot. He did apply for some scholarships, but found that most of the money had dried up with increased national financial problems in the U.S. He also used his Missouri A+ scholarship for OTC, majoring in Graphic Design. After graduation, he will look for a job in graphic design, specializing in t-shirts. He still works in the family farm business and in his spare time, helps Paul with the Model T restoration.

Lloyd Oberbeck – Lloyd was accepted to the University of Missouri and used his Missouri A+ scholarship plus additional scholarships to start a pre-med program. The summer after Fred's victory, he spent in Prague, and the next year in Rome. He found travel so stimulating that he changed his major to Hospitality Management.

Grant Rumfelt – Grant worked during the semester following high school graduation. He then attended College of the Ozarks for one semester, then transferred to OTC. He currently attends Missouri State University in Springfield, MO. After graduation, he plans to apply to the Missouri Highway Patrol.

Shane Sell – Shane received a scholarship, and attended Central Methodist College for a year, then transferred to MSU in Springfield, MO, majoring in majoring in Computer Information Systems with a minor in Computer Science. He plans to be a network administrator.

Ceira Gisselbeck - Ceira used her Missouri A + scholarship for OTC, majoring in Biology. She worked at a Dollar Store her first year, and was quickly promoted to Assistant Manager. The summer after her second year of college, she married Broghan Fields and moved to North Carolina to continue school. When Broghan was deployed to Kyrgistan, Ceira moved back to the Springfield area to continue college. She worked as a Direct Support Professional for people with developmental disabilities and has changed her focus. She plans to continue coursework toward a goal as a nutritionist/dietician/lifetime wellness expert.

Broghan Fields – Broghan enlisted in the Marines right out of high school, and specialized as a machine gunner. After one year in North Carolina, he proposed to Ceira. They married and moved to North Carolina. After his deployment in the Middle East, he returned to North Carolina in November, 2013. He does not plan to re-enlist, and will apply for college with a major in Psychology.

Henry Stratmann – The year after he programmed the national champion robot, Henry became a celebrity of sorts within the MSU Computer Sciences majors. He was frequently called on to speak about his experiences to high school students interested in MSU. He graduated with a BS degree in Computer Information Systems and accepted a position with a company in Kansas City, Missouri that specializes in developing software for hospitals and physicians' offices.

Bob Gibson – Gibby retired from Conway High School the year after Paul and Phillip graduated. During the Fred year, he earned about $1,000 for his 500 hours of extra-duty work supervising robotics. The next year, he received no extra funds due to state cutbacks and at the end of the school year, retired from the Missouri School System. He taught the next year at Akiak, Alaska. He is currently teaching in Omaha, Arkansas where he hopes to stay for five years and become invested in the Arkansas Teacher Retirement program.

Billy Coyle – The fall following the national robotics championship, he donated $50,000 to start a foundation for Conway, Missouri high school students to help with college tuition. He continues to teach Science part-time because he loves to see students learn.

BRAGGING RIGHTS

⊹

Each of the high school seniors who were on the Fred driver/spotter team - Shane Sell, Grant Rumfelt, Lloyd Oberbeck - had something unique for their college freshman meet and greet time. When asked to talk about a highlight of high school, no one could top, "I was part of the first high school robotics team to win a national championship."

The next year, BEST had an international championship when Canada joined the event. CHS came in 20th, no small feat for a small, rural U.S. school.

Reunion - Phillip, Ceira & Paul

SMALL SCHOOL REALITIES

After Mr. Gibson left Conway High School, no one took over robotics sponsorship, so the club disbanded. Paul's younger sister and others who were apprentices in robotics during the Fred and Brad years hope that changes before they graduate high school.

The shop teacher left the district, and with tech teachers in short supply, the school board decided to sell off the equipment, so no tech programs currently exist at Conway, Missouri in the Laclede County R-I school district. Tech students now travel to Lebanon, Missouri for classes in the Laclede County R-III school district.

In the U.S., local property taxes support the basics of public school operations, accounting for about half the total funds available to school districts each year. Districts that lie in rural and small town areas, such as Laclede County R-I in Missouri, have far less funds for daily operations that do districts with large industrial operations within their tax base.

THE BEST AWARD – GENERAL RULES

The BEST Award is presented to the team that best embodies the concept of *Boosting Engineering, Science, and Technology*. This concept recognizes that inclusiveness, diversity of participation, exposure to and use of the engineering process, sportsmanship, teamwork, creativity, positive attitude and enthusiasm, and school and community involvement play significant roles in a team's competitive experience and contribute to student success in the competition beyond winning an award.

In accordance with the BEST philosophy, **materials submitted by teams must be the work of students.** The involvement of student peers in auxiliary roles to support a school's official BEST team with the documentation – i.e., journalists, photographers, artists, musicians – is encouraged.

Best Award Judging Evaluation and Criteria Evaluation of competitors will be based on the criteria outlined in these guidelines. An evaluation score of a total possible 100 points will be composed of the following:

•Category I - Project Engineering Notebook (mandatory for ALL teams, including teams NOT competing in the BEST Award)

• Category II - Marketing Presentation (at hub's discretion for BEST Award inclusion)

• Category III - Team Exhibit and Interviews (at hub's discretion for BEST Award inclusion)

• Category IV - Spirit and Sportsmanship (mandatory for all BEST Award teams)

• Category V - Robot Performance (mandatory for all BEST Award teams)

Hubs are required to judge at least four of the above five categories using one of the following scenarios:

Scenario 1: (preferred)

Judging Category	Point Value
Project Engineering Notebook	25 points
Marketing Presentation	25 points
Team Exhibit and Interviews	20 points
Spirit and Sportsmanship	15 points
Robot Performance	15 points
Total	**100 points**

Scenario 2:

Judging Category	Point Value
Project Engineering Notebook	25 points
Marketing Presentation	25 points
Spirit and Sportsmanship	15 points
Robot Performance	15 points
Total	**80 points**

Scenario 3:

Judging Category	Point Value
Project Engineering Notebook	25 points
Team Exhibit and Interviews	20 points
Spirit and Sportsmanship	15 points
Robot Performance	15 points
Total	• **75 points**

Judging Procedure
- A distinguished team of judges from private and public sectors with technical and non-technical expertise will evaluate teams. Judges will serve on a rotation schedule.
- As each team completes a category, it will be assigned a category score that is the average of individual scores of the judges reviewing it.
- Teams should know in advance that scores among many teams frequently differ by only fractions of a point.
- All teams do not have equal resources. The judges may take into consideration the resources available to teams to conduct their BEST programs (financial or technology resources, for example) so that a team is not penalized for its limited resources.

Judging Results
- Each advancing team will be mailed a copy of its score sheets following their local competition. Score sheets of non-advancing teams will be mailed upon request.
- Teams advancing to the Regional championships can use judges' comments to make improvements as they wish.

BEST no longer sponsors a national or international robotics championship, citing costs to schools as prohibitive.

No one will ever match the feat accomplished by Team Fred from Conway, Missouri.

CITY LIMIT
Conway
POP 743

2010 National High School
Robotics Champions

New City Limit Sign

ABOUT THE AUTHOR

Joyce Ragland is the author of more than one hundred academic publications including two books and various professional articles, in addition to short stories and poetry. Her latest book, *Dread the FRED*, takes her back to her Conway Missouri high school alma mater. Ragland holds a BA in Music; MA and Ed.D. in Education-Administration. After a career as teacher, professor, assistant dean, she is devoting time to her lifelong love of creative writing. Joyce is founder and president of the Ella Ragland Art charity and raises funds for local Alzheimer patient projects. She resides in Springfield, Missouri with a companion, Bessie Jo, a short haired Border Collie.

www.ingramcontent.com/pod-product-compliance
Lightning Source LLC
Chambersburg PA
CBHW060757050426
42449CB00008B/1437